HDTV For Dummies®, 2nd Edit

Cheat Sheet

HDTV Technology Compariso

HDTV Type	Pros	
Plasma	Thin, excellent picture, larger screen sizes in flat panel category	Expensive, slightly prone to burn-in, only a handful of 1080p models
LCD flat panel	Thin, excellent picture, capable of 1080p	Expensive but declining, so-so black reproduction
CRT direct-view	Great colors and blacks, cheap	Bulky, smallest screen size, lower resolution
LCD projection	Excellent picture, thin cabinet	So-so black reproduction, expensive compared to CRT projection, only a handful of 1080p models
DLP projection	Excellent picture, thin cabinet, capable of 1080p	Expensive compared to CRT projection, rainbow effect noticed by small number of viewers on single chip color wheel models
LCoS projection	Capable of 1080p, great picture	Expensive compared to CRT projection, only a few models
CRT front projection	Potentially the finest picture available	Typically expensive, fussy to set up, limited availability
CRT rear projection	Best screen size:price ratio	Very bulky, requires convergence adjustment, prone to burn-in, limited availability

HDTV Cabinet Comparison

Cabinet Type	Pros	Cons
Flat panel	Thin, sexy, can be hung on wall	Most expensive for screen size
Microprojector* rear projection	Relatively thin, cheaper than flat panel	More expensive than CRT rear projection
Rear projection	Big screen	Bulky cabinet
Front projection	Biggest screen	Most expensive, requires more set up
Direct view	Cheapest	Smallest screen, bulky in terms of depth

** Microprojector technologies include LCD, DLP and LCoS*

For Dummies: Bestselling Book Series for Beginners

HDTV For Dummies®, 2nd Edition

Digital TV Format Comparison

Format	Resolution	Progressive Scan	Comments
1080p	1,920 x 1,080	Yes	Highest resolution, to be available on HD-DVD and Blu-ray disc players
1080i	1,920 x 1,080	No	Highest broadcast resolution signal, but interlaced, so not as good as 720p on rapidly moving images
720p	1,280 x 720	Yes	The threshold for true HDTV, combines high resolution with progressive scan
480p	720 x 480	Yes	Resolution only slightly higher than analog TV, but progressive scan and widescreen available on anamorphic DVDs
480i	640 x 480	No	Equivalent to analog TV, interlaced and not widescreen

Looking for the Logos

When you're shopping for an HDTV, you find that the folks at the Consumer Electronics Association (CEA — `www.ce.org/hdtv`) have come up with some handy dandy logos that can help you quickly identify and understand the HDTV capabilities of any TV set you're considering. The Web site has even more logos, but here are the ones you're most likely to see.

Logo	What It Means
	These TVs are fully integrated HDTVs — in other words, they can display HDTV resolutions, in a true 16:9 widescreen mode, without requiring any kind of external tuner or set-top box.
	These HDTV-ready TVs can handle true HDTV signals (that is, 720p or 1080i) with full resolution and widescreen displays, but require an external HDTV tuner or cable/satellite set-top box to tune into broadcasts.
	These external tuners let you tune into HDTV broadcasts when using an HDTV-ready or HDTV monitor. If your TV has an HDTV logo, you don't need an external tuner for picking up broadcast HDTV signals.
	These TVs are widescreen (16:9) and have higher resolutions than old-fashioned analog TVs, but can't fully reproduce HDTV. There's nothing wrong with EDTVs (in fact, they're great for DVD watching), but they aren't HDTVs.
	These televisions are capable of tuning into digital TV broadcasts, but they display such programs at the lower resolutions of today's analog TVs. SDTVs let you take advantage of digital television at a lower cost, without the benefits of HDTV.
	These are external tuners designed to receive digital TV broadcasts and decode them into standard-definition TV signals. SDTV tuners will become useful in the future when broadcasters switch to digital TV broadcasts only and turn off their existing analog broadcasts — an inexpensive SDTV tuner will let you pick up those broadcasts and watch them on your old analog TV.

HDTV FOR DUMMIES®

2ND EDITION

by Danny Briere and Pat Hurley

BICENTENNIAL 1807 WILEY 2007 BICENTENNIAL

Wiley Publishing, Inc.

HDTV For Dummies®, 2nd Edition

Published by
Wiley Publishing, Inc.
111 River Street
Hoboken, NJ 07030-5774

www.wiley.com

Published by Wiley Publishing, Inc., Indianapolis, Indiana

Published simultaneously in Canada

For general information on our other products and services, please contact our Customer Care Department within the U.S. at 800-762-2974, outside the U.S. at 317-572-3993, or fax 317-572-4002.

For technical support, please visit www.wiley.com/techsupport.

Wiley also publishes its books in a variety of electronic formats. Some content that appears in print may not be available in electronic books.

Library of Congress Control Number: 2006936828

ISBN: 978-0-470-09673-4

Manufactured in the United States of America

10 9 8 7 6 5 4 3 2

2O/SQ/RS/QW/IN

About the Author

Danny Briere founded TeleChoice, Inc., a telecommunications consulting company, in 1985 and now serves as CEO of the company. Widely known throughout the telecommunications and networking industry, Danny has written more than 1,000 articles about telecommunications topics and has authored or edited ten books, including *Smart Homes For Dummies* (now in its second edition), *Wireless Home Networking For Dummies* (now in its second edition), *Wireless Hacks and Mods For Dummies, Windows XP Media Center Edition 2004 PC For Dummies, and Home Theater For Dummies* (now in its second edition). He is frequently quoted by leading publications on telecommunications and technology topics and can often be seen on major TV networks, providing analysis on the latest communications news and breakthroughs. Danny splits his time between Mansfield Center, Connecticut, and his island home on Great Diamond Island, ME, with his wife and four children.

About the Author

Pat Hurley is Director of Research with TeleChoice, Inc., specializing in emerging telecommunications and digital home technologies, particularly all the latest consumer electronics, access gear, and home technologies, including wireless LANs, DSL, cable modems, satellite services, and home-networking services. Pat frequently consults with the leading telecommunications carriers, equipment vendors, consumer goods manufacturers, and other players in the telecommunications and consumer electronics industries. Pat is the co-author of *Smart Homes For Dummies, Wireless Home Networking For Dummies, Wireless Hacks and Mods for Dummies, Windows XP Media Center Edition 2004 PC For Dummies,* and *Home Theater For Dummies.* He lives in San Diego, California, with his wife, a fiery red-headed toddler named Annabel, and two smelly dogs.

Authors' Acknowledgments

Danny wants to thank his wife for not freaking out when he's gotten big TV sets and lots of equipment, and for accepting his bribes of absolutely everything he could think of to buy the extra time to research this book. She's a trooper. However, this "Thank You" is not about her, but about co-author Pat, who's been so much fun to work with on these *For Dummies* books over the years. Although Pat lives on the other coast of the U.S., and the authors rarely actually see each other, they work side by side through videoconferencing, IM, and e-mail, and (oh, yes) voice calls. Danny has heard Pat's toddler's voice and wife's pleas for dinner in the background, and feels he knows Pat's house by the location of the different devices he so often writes about ("I'm IMing you from the Media Center PC in the living room . . ."). Pat's wife is used to Danny's calls after hours and on weekends and has been nice enough to let those slip by — well, at least occasionally — and Danny thanks her immensely for never chewing him out on the phone. (Danny's sure that she chews out Pat instead, but Pat's too Navy to tell his superior about that.) When Pat went in the hospital in the early stages of the first edition of this book, it simply was not as much fun writing a book without Pat's incessant rantings about how stupid people are in the marketing of their products and about how nobody's Web site is worth a hoot. (Being a good former Navy man, he actually never says *hoot*.) These books are always a pain in the ruckus to do, but are made all the more pleasant with a co-writer who has a sense of humor (or the smarts to laugh at his boss's bad jokes, whichever applies).

Pat dedicates this book to his daughter, Annabel. Yeah, little girl, that big plasma is all for you (and for Thomas the Tank Engine, and Percy, and Toby and Gordon and Duck and all your other favorite "useful engines"). Now if only they'd get all of these shows in nice 1080p HDTV! Pat thanks his wife Christine, who again gave more support to him during this process than he deserved, but definitely all that he needed.

Many folks in the HDTV industry gave us assistance as we researched this book, including Sara Schiffler at Chief Manufacturing; Dan McCarron and Chelsea Vander Groef at JVC; Mark Knox at Toshiba; Eric Smith and Joelle Kenealey at Control4; Carl Cranbill and Michelle Fox at MovieBeam; Jim Denney and Krista Wierzbici at TiVo; Tim Kelly, Will Rosser and Corey Vasquez at DISH Network; and Robert Mercer at DIRECTV. Much thanks to all of you. We would also like to thank Jenny Miller at the Consumer Electronics Association (CEA), for her assistance and guidance.

Publisher's Acknowledgments

We're proud of this book; please send us your comments through our online registration form located at `www.dummies.com/register/`.

Some of the people who helped bring this book to market include the following:

Acquisitions, Editorial, and Media Development

Project Editor: Rebecca Senninger

(Previous Edition: Pat O'Brien)

Acquisitions Editor: Melody Layne

Copy Editor: Virginia Sanders

Technical Editor: Dale Cripps

Editorial Manager: Leah Cameron

Media Development Manager: Laura VanWinkle

Editorial Assistant: Amanda Foxworth

Sr. Editorial Assistant: Cherie Case

Cartoons: Rich Tennant (`www.the5thwave.com`)

Composition Services

Project Coordinator: Adrienne Martinez

Layout and Graphics: Joyce Haughey, Stephanie D. Jumper, Barry Offringa, Laura Pence, Heather Ryan

Proofreaders: Cynthia Fields, Jessica Kramer, Techbooks

Indexer: Techbooks

Anniversary Logo Design: Richard Pacifico

Publishing and Editorial for Technology Dummies

Richard Swadley, Vice President and Executive Group Publisher

Andy Cummings, Vice President and Publisher

Mary Bednarek, Executive Acquisitions Director

Mary C. Corder, Editorial Director

Publishing for Consumer Dummies

Diane Graves Steele, Vice President and Publisher

Joyce Pepple, Acquisitions Director

Composition Services

Gerry Fahey, Vice President of Production Services

Debbie Stailey, Director of Composition Services

Contents at a Glance

Cartoons at a Glance

By Rich Tennant

"Darn it, I wish they'd sent one that already has the HDTV tuner built in."

page 7

page 79

"The picture is so sharp, you can see the Hot Wheels logo on the cars that the monster is throwing across the city."

page 247

The 5th Wave By Rich Tennant

WARREN ENJOYED SHOWING OFF HIS NEW 52" FLAT PANEL HDTV DINNER

Dot's Veal Cutlet Peas and Yams

page 129

page 287

"There's a slight pause here where the PS2 sequences your DNA to determine your preferences."

page 167

"HD TV, huh? Well'p there goes the ambiance."

page 209

Fax: 978-546-7747
E-mail: richtennant@the5thwave.com
World Wide Web: www.the5thwave.com

Table of Contents

Part II: Getting HDTV Programming........................... 79

Introduction

Welcome to *HDTV For Dummies,* 2nd Edition. HDTV is the hottest technology to hit your local electronics store since the advent of cell phones. HDTVs are getting bigger, better, cheaper, more sophisticated, and more useful every day. Because you've bought this book, we figure that not only do you agree with us, you're already part of the HDTV movement. To arms, comrade!

One of the most appealing things about the current crop of HDTVs is the ease with which you can set up an HDTV-powered home theater, including surround sound and an awesome picture. However, the rapidly dropping price of HDTVs might be the most attractive aspect of all — we've reached a point in time where you don't have to be rich to consider an HDTV for the bedroom, too! Figuring out which HDTV to buy can be confusing because you can find all sorts of technologies, sizes, and standards. And then, after you decide what to buy, making sure it works with all your other gear — your DVD, camcorder, VCR, set-top box, and so on — can be even more confusing. That's where this book comes in handy. Our aim is to make sure you get the most bang for your buck (or franc, or peso, or whatever — even euros!).

About This Book

If you're thinking of purchasing an HDTV and installing it in your home, this book is for you. Even if you've already purchased the HDTV itself, this book helps you install and configure the HDTV. What's more, this book helps you get the most out of your investment after it's up and running.

With this book in hand, you have all the information that you need to know about the following topics:

- Planning your HDTV system, including all sorts of accessories
- Evaluating and selecting the right HDTV for your home
- Installing and configuring the HDTV equipment in your home
- Hooking up your HDTV to the right high-definition programming sources
- Adding A/V entertainment gear and accessories to your HDTV

- Playing video games in high-def splendor
- Finding content for your HDTV on the Internet
- Enhancing your HDTV environment so you can have your own HDTV theater

Foolish Assumptions

While writing this book, we had the perfect excuse to watch a lot of TV. ("Not now, honey, I'm working on my book.") It's been great.

Still, we already know what we like and what we value in an HDTV. But we're writing this book for you! While writing, we pondered all kinds of questions concerning our readers. Who are you? Where are you? What did you eat for lunch? Which movies tweak your interest? How do your HDTV desires line up with your budget? Queries like that fill our minds constantly, much to the consternation of our spouses, who prefer that we think more useful thoughts, like "Shouldn't I take out the trash or empty the Diaper Genie?"

Because we never got to meet you in person, we ended up making a few assumptions about you and what you want from this book. Here's a peek at our thoughts about you:

- You love movies, television shows, or video games — or perhaps all three.
- You've experienced widescreens and surround sound at the theater, and you liked it.
- For one reason or another, a 19-inch TV set with a single built-in speaker doesn't adequately meet your audio or video entertainment needs.
- You probably own a computer, or will soon.
- You don't shy away from high-tech products, but you also aren't the first person on the block with the latest electronic goodie.
- The weird technicalities of your A/V system make you dizzier than a Marilyn Monroe movie.
- You know something about the Internet (getting online, preferably with a broadband service) and about how to navigate the Web.
- You, or someone in your family, enjoy watching movies, listening to MP3 audio, playing games, and possibly making movies on your computer.

If that describes you today, in a prior life, or in another personality, this book is for you.

How This Book Is Organized

This book is organized into several chapters that are grouped into seven parts. The chapters are presented in a logical order — flowing from purchasing your HDTV, through all the things you'd want to hook up to it to exploit its very existence, to some detailed drill-down discussions about high-definition topics that help you get the most out of your HDTV environment. However, you can feel free to use the book as a reference and read the chapters in any order that you want. We wrote it that way.

Part I: HDTV Fundamentals

The first part of the book is a primer on HDTV. If you've never owned an advanced-level television — much less attempted to install one — check out this part of the book to find all the background information and techno-geek lingo that you need to feel comfortable. Chapter 1 presents general HDTV concepts. Chapter 2 discusses the most popular HDTV technologies and familiarizes you with high-definition terminology, and also provides guidance on making buying decisions. Chapter 3 introduces you to several popular ports, interfaces, jacks, plugs, cables and the sort — everything you need to know about connecting your HDTV into your existing audio/visual environment. Chapters 4 and 5 dig into the details of unpacking, installing, and mounting your HDTV.

Part II: Getting HDTV Programming

Part II discusses all the different forms of high-definition signals that you can access and/or subscribe to, in order to really take advantage of your great new investment. In Chapter 6, we talk at a high level about what is available in high-definition format now, and what's coming in the near future. Then in the next three chapters, we dive into each option — over-the-air broadcasts, cable, and satellite — so you know where you can get what signals and what to expect from each kind of service. Then we show you what the Internet has in store for you. Internet-based content is growing rapidly and looks fabulous on that new HDTV of yours.

Part III: Movie Machines

The broadcast programming we discuss in Part II is nice, but you want to watch a lot of other content, too — we're talking about all those DVDs and

VHS movies you own. In this part, we talk about the complexities of interfacing your DVD player/recorders and VHS VCRs with your HDTV. We also delve into the exciting world of digital video recorders — DVRs or TiVos, as some people call them (referring to one brand on the market generically). With these devices, you can record all sorts of content for later watching.

Part IV: HDTV Gadgets Galore

After you get your HDTV system installed and running, you certainly want to use it for even neater things if you can. Part IV of the book presents many cool things that you can do with your HDTV, including playing multi-user computer games, connecting your camcorder to preview your future *America's Funniest Home Videos* submission, and operating various types of smart home conveniences from the luxury of your bedroom.

Part V: Sensory Overload

In this part, we spend a lot of time drilling down in detail about the nitty gritty of getting your home HDTV viewing experience as good as you can get it. We start with an extensive discussion about audio basics and how they affect your HDTV viewing experience. We then discuss built-in speaker options versus external sound systems, and the advantages of a surround-sound-powered HDTV experience. We also tell you how to use lighting, room treatments, and other nuances in your home to create a true HDTV theater. Finally, we show you how to enhance your HDTV so that it's tuned perfectly for your use, and we also tell you about a range of little "black boxes" that can help you optimize all of the non-HDTV signals you send to your HDTV, making them look better on your HDTV big screen.

Part VI: Geek Stuff

Of course, the more you know, the geekier these topics start to appear. Before long, you might be dreaming of ultra-high-tech ways to expand your system's capabilities. This part of the book unashamedly encourages that bad habit. First, we delve into the details of your TV picture, and then we cover the various ways you can accomplish HDTV, including front and rear projection, plasma and LCD screens, and the good old CRT approach. When you finish this part, you should know much more than the average salesperson walking the show floor at your local TV store.

Part VII: The Part of Tens

Part VII provides two top-ten lists that we think you'll find interesting: ten places to look online and locally to buy an HDTV (and at the same time, find out more about your options), and ten devices to connect to your HDTV.

Icons Used in This Book

Everyone these days is hyper-busy, with no time to waste. To help you find the especially useful nuggets of information in this book, we've marked the information with little icons in the margin. Here are the icons we use in this book:

As you can probably guess, the Tip icon calls your attention to information that saves you time or maybe even money. If your time is really crunched, you might try just skimming through the book and reading the tips.

The little bomb in the margin alerts you to pay close attention and tread softly. You don't want to waste time or money fixing a problem that could have been avoided in the first place.

This icon is your clue to take special note of the advice that you find there . . . or note that this paragraph reinforces information that has been provided elsewhere in the book. Bottom line: You can accomplish the task more effectively if you remember this information.

Face it, HDTVs and home entertainment systems are high-tech toys that make use of some pretty complicated technology. For the most part, however, you don't need to know how it all works. The Technical Stuff icon identifies the paragraphs that you can simply skip if you're in a hurry or you just don't care to know.

Where to Go from Here

Where you go next in this book depends on where you are in the process of planning, buying, installing, configuring, and/or using your HDTV. If HDTV in particular is totally new to you, we recommend that you start at the beginning with Part I. If you feel comfortable with HDTVs and all of the connections, you might just read Chapter 2 about buying advice. If you already have your HDTV but need to mount it, check out our great installation and mounting help in Chapters 4 and 5. Already installed? You might want to check that it's optimally configured with the chapters in Part V. If your HDTV is installed and you want to know more about what you can connect to it, Parts II, III, and IV all talk about neat things you can channel (oops, pun) to your HDTV. If you're in the depths of analyzing your equipment options, Part VI might be the best place to find the details you are looking for. There's simply a lot here for whatever you need to know about HDTV.

Either way, happy reading!

Part I
HDTV Fundamentals

The 5th Wave By Rich Tennant

"Darn it, I wish they'd sent one that already has the HDTV tuner built in."

In this part . . .

If you ever had the exciting opportunity to go to FAO Schwartz's flagship store in New York City before it closed, you were greeted by a huge, fanciful clock. The song played by the clock is very appropriate here as we begin to talk about HDTVs — a chiming "Welcome to our world, welcome to our world, welcome to our world of toys!"

Oh boy, are HDTVs fun — and that fun, in Part I, is just beginning. We introduce you to the world of HDTV, our world of toys. In Chapter 1, we introduce you to the key acronym of the HDTV world, ATSC, and tell you why you should care about it. We explain the foundations of HDTV, of things like resolution, scan types, and aspect ratios. We talk of 480i, 480p, 720p and 1080i, which sound more like something from *I, Robot* than from Circuit City.

We also help you go shopping. In Chapter 2, we cover the key buying criteria for HDTVs and how to best match a TV to your needs, environment, existing audio/video gear, and other HDTV decision-affecting facets of your life.

We continue our introduction by talking about the backs, sides, tops, and other parts of HDTV systems — all the places where you'd connect to your HDTV display other sources and gear, such as DVDs, VCRs, camcorders, satellite receivers, and more.

We finish off your intro to HDTV by delving into how to set it up (including how to mount it if you need to mount it on the wall or ceiling) and how to plug in all your myriad of cables. When you're done with this part, you should be all ready to take part in the world of high-definition TV, which we talk about in Part II.

So you're off on your HDTV adventure if you're starting with Part I right away. Be sure to turn off your toys when you're done playing with them!

Chapter 1

What the Heck Is HDTV?

In This Chapter

- Understanding the acronyms
- Transmitting from ATSC to the world
- Going wide
- Avoiding the pitfalls

Since the transition to color TV in the 1950s and '60s, nothing — nothing!! — has had as much impact on the TV world as HDTV (high-definition TV) and digital TV. That's right. TV is going digital, following in the footsteps of, well, everything.

We're in the early days of this transition to a digital TV world (a lot of TV programming is still all-analog, for example), and this stage of the game can be confusing. In this chapter, we alleviate HDTV anxiety by telling you what you need to know about HDTV, ATSC, DTV, and a bunch of other acronyms and tech terms. We also tell you *why* you'd want to know these terms and concepts, how great HDTV is, and what an improvement it is over today's analog TV (as you can see when you tune in to HDTV). Finally, we guide you through the confusing back alleys of HDTV and digital TV, making sure you know what's HDTV and what's not.

Almost everyone involved with HDTV has noticed that consumer interest is incredibly high with all things HDTV! As a result, a lot of device makers and other manufacturers are trying to cash in on the action by saying their products are "HDTV" (when they are not) or talking about such things as "HDTV-compatible" when it might be meaningless (like on a surge protector/electrical plug strip). Be on the lookout for such interlopers and insist on true HDTV functionality. We help you in this chapter — read on!

Oh, Say, Can You ATSC?

A long time ago (over 50 years ago — longer than even Danny has been alive!), in a galaxy far, far . . . errr, actually right here in the United States . . . a group called the *NTSC* (National Television System Committee) put together a group of technical specifications and standards that define television as we know it today. Sure, some changes have been made in those 50 years (such as the addition of color), but today's analog TVs are built on this NTSC system.

Fifty years is a long time for any technology to dominate. Indeed, technologies and components used in television-transmission systems, cameras, recording systems, and display systems (the TVs themselves) have long been capable of doing something more.

In the 1980s, the *ATSC* (Advanced Television System Committee) was formed to move TV forward. Many years later (1996), the FCC (Federal Communications Commission — the folks who set standards for TV broadcasts, regulate phone companies, and fine Howard Stern) adopted the ATSC's recommendations for a digital-television system. ATSC standards use newer-than-1953 technology to give you TV like you've never had before:

- Widescreen images like those in the movies
- Greater detail — up to six times more detail
- Sharper images
- Smoother, more film-like images with no video flicker
- All digital, with none of the *ghost* images — where you see a translucent version of the image on your screen, slightly offset — and other image problems found in analog TV

Please note that ATSC is not the only game in town when it comes to digital TV and HDTV. In most European countries and in many other parts of the world, the DVB (Digital Video Broadcasting) standard applies. (Terrestrial, cable, and satellite variants exist, noted by a -t, -c, or -s appended to the DVB.) In Japan and a few other countries, there's a system known as ISDB (Integrated Service Digital Broadcasting). These systems all use different mechanisms to *encode* (or digitize and compress) video signals for transport over the airwaves, over a cable, or via satellite. The impact here is that the device you use to tune in your HDTV signals differs depending upon your country.

Regardless of which system is in use where you live, the important thing to you as a viewer is the picture itself, and that's where HDTV is truly a universal phenomenon. HDTV pictures contain two to six times more picture detail than older analog *standard-definition* systems (again, this is true regardless of which country you're dealing with). This extra detail lets you display your TV content on a bigger screen and still see a great picture. You also see more vivid colors, a wider screen presentation, and the increased picture quality enabled by digital transmission. No more ghosts and snowed-out pictures — digital is usually either a great picture (most of the time) or essentially no picture at all (on those rare occasions when you don't have a good enough signal).

Throughout this book, we focus on the U.S.-based ATSC system, but our real focus is on the HDTV formats we're about to discuss, such as 720p and 1080i, which apply no matter where you live and no matter how your TV is delivered to you.

Powerful Performance

HDTV is all about giving you a bigger and better picture, better audio, and generally making your TV-watching experience more like a movie-watching experience. (Digital TV, or DTV, in general also does the same thing, but some digital TV variants are *not* high definition, and we discuss them in the following sections.) In fact, at its best, HDTV is so realistic that it's often described as "looking through a window" — as if you're really there, not just watching a program.

Video standards

In this section, we discuss the characteristics of the HDTV programming itself (be it broadcast over the airwaves, over cable, via satellite, or saved on a hard drive or optical disc). There's a related category of the *display* characteristics for HDTV — meaning what your HDTV can actually show you on the screen. Many HDTVs can accept different types of HDTV signals and then transform them into the resolution, aspect ratio, and other such formats that work best on the HDTV display itself. We talk in detail about this in Chapter 22.

You need to understand four essential concepts when comparing different video standards:

- **Resolution:** The number of individual picture elements that make up a TV image. The higher the resolution, the more detailed the image and the sharper the image.

  Resolution is defined by one of two factors:

  - *Lines:* The number of left-to-right lines — counted vertically, like a stack of pancakes — the TV can display. CRT-based TVs (tube TVs) are rated this way.
  - *Pixels:* The number of pixels across the screen times the number of pixels up and down. Fixed-pixel displays (plasmas, LCDs, DLPs, and the like) are rated this way.

- **Scan type** comes in two forms:
  - *Interlaced scan:* These TV images are created by lighting up every other row of horizontal lines on the screen in one instant and then going back through and lighting up the remainder of the lines in the next instant. It happens so fast that your eye can't really tell it's happening. In an interlaced system, these groups of lines (each consisting of half of the picture) are known as *fields.*
  - *Progressive scan:* These systems light all the horizontal lines in the same instant, which can make the image seem "smoother" and more like film (or real life). In progressive scan, this grouping of all the lines is called a *frame.* Two interlaced *fields* combined together equal one full *frame.*

- **Scan rate** is the measure of how often a picture is redrawn on the screen, measured in terms of the number of fields (for interlaced scan) or full frames (for progressive scan) that are drawn on the screen per second. In the United States, this is typically either 30 or 60 times per second (often called *hertz*). In European markets, it's often 50 hertz.

  Movies themselves are usually filmed at 24 frames per second, which then must be converted to 50 or 60 during the process of turning film into video.

- **Aspect ratio** (the shape of your TV picture):
  - Traditional TVs have a 4:3 *aspect ratio.* This means that for every 4 units of measure across the screen, you have 3 units of screen height. For example, if the screen is 12 inches wide, it's 9 inches high.
  - HDTVs have a 16:9 aspect ratio — which makes the screen relatively much wider for the same height, compared to a 4:3 TV. Most movies are widescreen (16:9, or even wider), so HDTVs can display most movies without the annoying *letterbox* black bars on the top and bottom of the screen. Figure 1-1 compares aspect ratios.

4:3 / 1.33:1 Standard TV and older movies	16:9 / 1.78:1 US Digital TV (HDTV)

Figure 1-1: Going widescreen with a 16:9 aspect ratio.

Don't get bogged down in up-front technical explanations of these concepts now. If you want to know all there is to know about such TV concepts as resolution, pixels, scan rates, and interlacing, run (don't walk) to Chapter 21 right now. We'll still be here when you come back.

HDTV standards

There isn't a single HDTV standard out there. Instead, digital TV systems contain dozens of different TV standards (with different resolutions, aspect ratios, and scan types and rates). Some of these standards are truly HDTV; most are not. In the real world, you deal with three primary formats that are considered true HDTV:

- **720p:** This provides 720 lines of resolution with progressive scan (hence the *p*). By comparison, NTSC has less than 480 lines of resolution. In pixel terms, it has a resolution of 1,280 across by 720 vertically. 720p uses a 16:9 widescreen aspect ratio. You can find 720p in HDTV broadcasts and also in recorded HDTV content like HD DVD and Blu-ray discs.
- **1080i:** This variant (the highest resolution within the ATSC standard) uses interlaced scanning but provides 1,080 lines of resolution. In pixel terms, 1080i fills your screen with 1,920 pixels across by 1,080 vertically. 1080i is also widescreen, with a 16:9 aspect ratio.
- **1080p:** The big dog in the HDTV world is 1080p, which provides the same number of pixels or lines as 1080i but does it in a progressive scan fashion, so all 1,920 x 1,080 of those picture elements are redrawn each time your screen is refreshed, rather than only half per refresh. Today, 1080p is found only in recorded HDTV formats such as Blu-ray and HD DVD.

If you see 720i listed as an option, don't believe it. Either it's a typo for 720p, or someone is trying to fool you. No broadcast standard permits 720 *interlaced* lines in a video frame at any frame rate.

True HDTV performance requires at least 720p performance. If a TV program, movie, or other content isn't at least 720p (either 720p or 1080i), it is *not* HDTV. If a TV can't display at least 720 lines of resolution, it is *not* HDTV-capable.

If a salesperson tries to tell you that an inexpensive plasma set, regular DVD, regular digital cable, or regular satellite TV is HDTV just because it's *digital,* it's not so.

Compatible DTV standards

720p, 1080i, and 1080p are the three main HDTV standards, but you can also find a lot of digital TV material that is broadcast at lower resolutions that don't quite make the grade as HDTV. You can still watch this programming on your HDTV. In fact, most HDTVs make this programming look better than it does on a regular TV, but remember: This stuff *is not* really HDTV:

- **480p (EDTV):** This *enhanced-definition* TV standard provides higher-than-NTSC resolution with progressive scan (NTSC is interlaced). EDTV can be (and often is) 16:9 widescreen, but it isn't required to be widescreen.
- **480i (SDTV):** This is interlaced, non-widescreen (4:3), standard-definition TV, equivalent to NTSC analog broadcasts.

Remember these different terms — HDTV, EDTV, and SDTV — when shopping. They often are in the product descriptions; you need to know exactly what you're buying.

Audio standards

The ATSC standard includes big improvements in the audio part of television — what you hear as part of any movie, video, or TV show. That's because ATSC includes Dolby Digital surround-sound capability in the overall standard for digital TV.

Dolby Digital (which we discuss in greater detail in Chapter 18) doesn't *always* mean surround sound. Some Dolby Digital soundtracks are stereo (two channels) or even mono (one channel). ATSC supports surround sound if a program's producer and broadcaster want to include it.

The NTSC broadcast standard supports only stereo audio (two channels) and not surround sound. Luckily, most DVDs (and some satellite and digital cable TV channels) include Dolby Digital soundtracks that can provide true surround sound. You can also use a home-theater receiver that supports systems like Dolby Pro Logic II (see Chapter 18) to create surround sound from these sources.

Dolby Digital and surround sound in general provide an audio soundtrack for TV shows and movies that — wait for it! — *surrounds* you and provides audio that matches the action on-screen. For example, surround sound might use speakers mounted in the rear of the room to reproduce ambient noises of the setting around the action, or it might give a 3D sense of space to those creepy footfalls of the bad guy sneaking up behind the protagonist.

Dolby Digital provides six channels (confusingly called *5.1*) of audio. Here's what they do:

- A center channel carries the dialogue being spoken by characters on your HDTV screen.
- Two main front channels handle left and right sound cues (and the soundtrack music) in stereo.
- Two surround channels (mounted in the rear of the room) provide a sense of 3D space.
- A Low-Frequency Effects (LFE) channel conveys deep bass sounds (such as exhausts rumbling and bombs exploding). The LFE channel is the ".1" in the 5.1 naming scheme for Dolby Digital. This channel gets a fraction rather than a whole number because it contains only low-frequency sounds, not sounds for the full range of human hearing.

Figure 1-2 shows a typical Dolby Digital surround-sound layout.

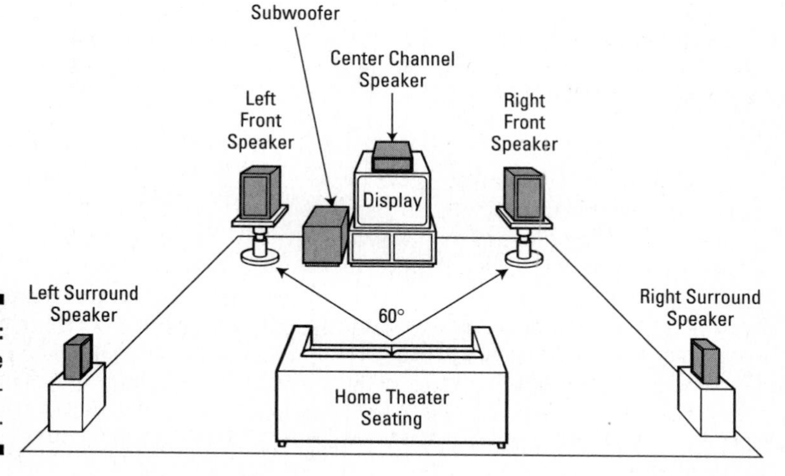

Figure 1-2: Doing the surround-sound thing.

We talk about surround sound in much more detail — including details on what sort of equipment you need to hear it properly in your HDTV viewing room — in Chapter 19.

Some new "better than Dolby Digital" standards are being included in Blu-ray or HD DVD disc players. These provide higher-quality audio and more surround channels. You can read about these new standards in Chapter 18 as well.

Perplexing Pitfalls

HDTV isn't the easiest thing in the world to get figured out — we've been dealing with it for years and still run into advertising and marketing mumbo-jumbo that make us say, "Huh?" The whole purpose of this book is to help you wade through the marketing manure and to get you up to speed on HDTV. So without further ado, here's a list of HDTV danger zones:

- **Digital confusion:** The biggest (and most prevalent) myth we see in the HDTV world is the notion that any kind of digital TV signal (such as digital cable, digital satellite, or DVD) is HDTV. This simply isn't true. A TV signal must be 720p resolution or higher to be considered high definition. We've seen too many people buy an HDTV, hook it up to their existing cable or satellite box, and then wonder why the picture isn't all that they'd imagined it would be — simply because they'd missed the step of activating an HDTV service to make it all work properly. In Part II, we go into detail about how to get HDTV broadcasts into your HDTV.

  Sometimes it's as easy as just changing the channel. Pat doesn't want to do this, but he's going to use the example of his dad and brother here. They had set up a new HDTV in Pat's dad's home and were watching a football game in what they thought was HDTV — and complaining how awful the picture was. It took about two seconds to change from the standard-definition broadcast on channel 4 to the high-definition one on channel 704. Jaws dropped, and all was well in the HDTV world again.

- **Input versus display resolutions:** When you're shopping for an HDTV, you can often see marketing and sales literature that includes a huge listing of resolutions that an HDTV can *accept.* The important thing to remember here is the difference between the resolution of the inputs (the source signals going into the HDTV) and the actual resolution of the picture on the screen. For example, an HDTV might say 480i/480p/720p/1080i on the box but have a display resolution of something like 1,280 x 720. What all these numbers mean is that you can tune into a program at any of these resolutions, and the TV converts the picture to the TV's display resolution. There's nothing wrong with this. (It's the standard behavior of just about every HDTV on the market, including all plasma, LCD, LCoS, and DLP TVs.) However, it leads to confusion when a consumer is buying an HDTV that's capable of 720p resolution but is convinced that it can display the full resolution of 1080i due to the confusing labeling on the box. One place to pay close attention to this phenomenon is in the case of EDTVs, discussed in the next bullet.

- **EDTV confusion:** EDTVs are TVs (typically 42 inches and under, plasma, flat-panel models) that cost a lot and can display progressive-scan images. However, they don't meet the minimum requirement of 720p, so they don't display true HDTV signals. Nothing is wrong with EDTVs; just don't be fooled into thinking you're getting an HDTV when you're not. We see a lot fewer EDTVs on the market today, but do beware: We can't think of a reason to buy an EDTV when HDTVs cost the same *or less* in most instances. Although some folks are of the opinion that given a certain viewing distance and screen size you really can't tell the difference, we'd still rather go with the HDTV variant.

- **Image scaling:** We're starting to see some new marketing being applied to an old concept — *image scalers* that can convert video signals from one resolution to another.

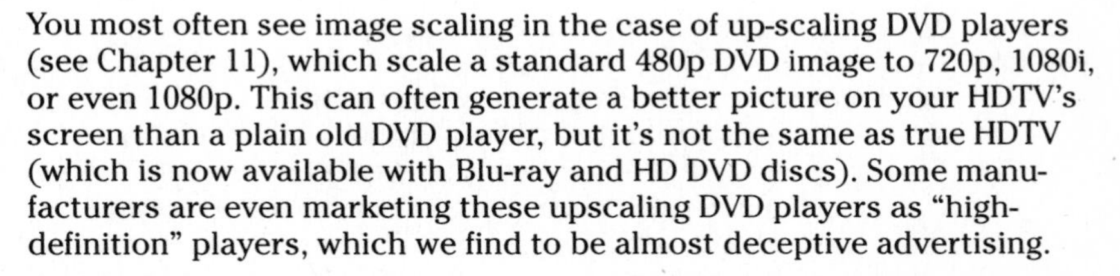

You most often see image scaling in the case of up-scaling DVD players (see Chapter 11), which scale a standard 480p DVD image to 720p, 1080i, or even 1080p. This can often generate a better picture on your HDTV's screen than a plain old DVD player, but it's not the same as true HDTV (which is now available with Blu-ray and HD DVD discs). Some manufacturers are even marketing these upscaling DVD players as "high-definition" players, which we find to be almost deceptive advertising.

- **The DTV tuner:** As HDTV (and DTV in general) becomes more prevalent, *DTV tuners* will become common. These tuners (discussed in Chapter 7) let older TVs "watch" DTV broadcasts. HDTV tuners do *not* turn older analog TVs into HDTVs. They just convert DTV signals to NTSC for display on an analog TV.

Chapter 2

Shopping Smart

In This Chapter

- Understanding your HDTV options
- Zeroing in on what's important in an HDTV
- Budgeting for your whole HDTV system
- Sizing your HDTV system
- Discerning the right feature set for your needs

Everyone has been there — you're standing in the electronics store looking at a wall of TVs, all tuned to the same channel, and they all pretty much look the same. So many TV sets, so little time, so hard to choose. So you pick the one on sale and leave, happy that you got "a deal." Been there, done that.

But no more. Our goal with *HDTV For Dummies,* 2nd Edition, is to make sure that you're no longer, well . . . dummies! Now you can be more educated. You can *know* that those TV sets are all misconfigured to appear a certain way in the bright lights of an electronics show floor. You can know to check how many digital interfaces the box has, how deep the chassis is, and how . . . well, lots of "hows."

Choosing the right HDTV for you is not the easiest thing to do. Heck, we wrote the book on it, and we still argue with each other about which HDTVs have the best bang for the buck. It depends on what you're trying to do, how much money you have, and what other A/V gear you have or intend to buy.

In this chapter, we walk you through a veritable buyer's guide to HDTVs — what to look for when shopping for just the right HDTV set for you. You can have too much HDTV (believe it or not) and the wrong type of TV for your intended use. Before you plunk down a lot of money on your well-earned HDTV surprise, make sure you're the best-informed buyer out there. Read on.

As we write this, HDTV prices are dropping more than 30 percent a year. Forty-two-inch plasma HDTVs — a deal at $3,000 in 2004 — are dropping near the $1,000 mark. You're getting more for your money than ever before. Part of deciding which HDTV to buy depends on how long you intend to use it and where you're putting it. For instance, although you might have four TVs in your house, you want to upgrade the living room first. However, as the price rate decreases, you figure you'll buy a new living room TV in a couple of years and move the present one to the bedroom. Makes sense to us. So with that in mind, do you really need the biggest, baddest, and hippest TV on the market? Why not get a TV today that fits well in your bedroom two years from now and wait for the better TVs to come down in price? It's up to you, obviously, but just don't feel like you need to satiate your TV diet upfront with a massive expenditure. You can get a lot of TV today for quite a little.

The 50,000-Foot View of HDTV

When you're trying to pick out the right HDTV for your needs, the available products break down into three major product groups, distinguished from each other by their display technology and cabinet type. By comparing display technology and cabinet type to your needs, you can easily rule out a whole bunch of different TVs and home in on the likely best ones for you.

HDTVs come in all sorts of different sizes and shapes. Some are flat panels that you can hang on the wall; others are projection systems like what you'd find in a movie theater. And, of course, some HDTVs are based on tubes that look just the way TVs have for decades (only with a better picture).

Each form of HDTV has some advantages and disadvantages. In Chapters 21 through 24, we discuss these pros and cons in much more detail, but here we give you just a quick overview to help you on your way. Skip ahead if you need the details and supporting information.

Flat-panel HDTVs

Flat-panel TVs — the super-thin models that you can literally hang on the wall — are the sexiest HDTVs available. These are the ones you see on *MTV Cribs* and that you might install in your tricked-out Escalade (yeah right). They're also good HDTVs. There are two main display technologies for flat-panel HDTVs:

- **Plasma:** These are the biggest flat-screens available, using a layer of gas trapped between two glass screens to create their images.
    - *Pros:* They're thin and sexy, have good picture and color, and are big. (You can even get displays greater than 100 inches — if you're willing to pay many tens of thousands of dollars for the privilege.)

- *Cons:* Not all are HDTV. They have less-than-perfect black, experience screen *burn-in* (where images like stock tickers are permanently displayed on your screen), and are costly.

  We mention *blacks* here. We're talking about how well an HDTV screen can reproduce dark tones and scenes on-screen — how well it creates black rather than gray colors.

- **LCD:** These flat-panel TVs use *liquid crystal displays,* just like those used in laptop computers.
  - *Pros:* They're the same as plasma, plus no burn-in.
  - *Cons:* The black is poorest, they're costly, and the angle of view is not as wide as it is with plasmas — meaning you'll need to be sitting closer to perpendicular to the screen for the best picture.

Projection HDTVs

Projection HDTVs project their picture from a smaller image source (either three small picture tubes or a digital system known as a *microprojector*) onto a screen. The screen can be either part of the HDTV itself (rear projection) or a separate screen hung on your wall (front projection).

- **Front-projection HDTV:** These are the HDTV equivalents to movie theater projectors, with a big screen on the wall and a separate projector mounted somewhere across the room.
  - *Pros:* They have the biggest screen, potentially the best picture, and can be portable.
  - *Cons:* They're expensive, complicated, and require setup/focus/maintenance.
- **Rear-projection HDTV:** The picture is projected on the back of a screen that is built into the HDTV itself.
  - *Pros:* You get the best bargain, no burn-in with microprojectors, and near flat-panel thinness for a microprojector.
  - *Cons:* You face burn-in for CRT (see the following section), expense for the microprojector, CRT is bulky, and a limited viewing angle in many models. You can't sit far off of perpendicular to the screen and still see a good picture — although some of the latest microprojector models have decent viewing angles.

CRT HDTVs

The final category of HDTVs is based on the good, old-fashioned picture tube — also known as the CRT, or *cathode-ray tube.*

- *Pros:* They're the cheapest HDTVs and they have great color and great blacks.
- *Cons:* They have a small screen, are bulky, and the resolution is lower than with digital displays.

What's Important in an HDTV

When looking at HDTVs, the following are the most important buying criteria for your purchase:

- **What's your budget?** We don't mean just for the TV set, but also for any attached home-theater surround-sound system, special remote controls, automated drapes, lighting controls, popcorn poppers, and the like. It makes a big difference if you're building a home HDTV theater or just putting a TV on the bureau in the bedroom.

  Just buying an HDTV is not enough. If you want to connect your DVD player to your HDTV, you might need additional cables and such. It's never as easy as just buying the HDTV and that's it. If you're really constrained on your budget, try hard to think of the details before you bring home that monster TV only to scrap around for cheap connection cables that give you a big but ugly picture.
- **What size do you need?** No, bigger is not always better. You can have a TV that's too large for your space or too small for your usage. There is an optimal range based on where you intend to place the TV and where you intend to sit.

  These first two items — size and budget — do a lot to narrow your choices before you get to any of the technical or usage criteria, so they're important to nail down first. If you want to fill an 8-foot wall with an image, unless you have a bank account the size of Bill Gates's, you're not going to do that with anything but a front projection system.
- **What do you plan to do with it?** Are you going to be watching a lot of sports events or movies? Playing video games? Believe it or not, certain types of HDTVs are better with certain types of content. Sports fanatics find a big, bright DLP projection system better for their tastes, everything else being equal, whereas people who watch CNN all day want to avoid plasma screen displays in a big way, due to the burn-in effects of static images (more on this in Part V).

Burn-in problems have been greatly reduced on today's plasma sets, but if you're going to have some sort of static image on-screen for hours at a time, it can still be an issue. At the very least, choose a model with special features like screensavers or screen "cleaning" anti-burn-in modes to accommodate your viewing.

- **What will you hook up to it?** If you already have a decent investment in A/V gear, that gear might dictate certain types (and numbers!) of interfaces or ports on your HDTV system, like these:
  - If you have an entertainment system designed around centralized video switching — using a receiver to switch among video sources and destinations — you need a receiver that can switch HDTV content. That might mean a new receiver, which can be pricey and cut into your budget.
  - Do you need a tuner or just an HDTV-ready display? Or will you be using a cable set-top box or satellite receiver to pick up HDTV signals?
- **What neat features do you want?** It's easy to be swayed by neat features, but in lots of implementations, you can't access them for various reasons. For instance, if you set up your system so all your signals come in over one cable connection, you might not be able to use your TV's dual-channel features. You could rely on your cable or satellite box for that. (We talk about these issues in Chapter 4.) Still, features are important to everyone, and we tell you in this chapter about which ones are the most important.

Budgeting for HDTV

If everyone had unlimited funds, everyone would simply buy the best of everything. That's why a market exists for super high-end gear — those with the big bucks often just buy the top of the line all the time because they can (and because they hire consultants to tell them that).

You're more likely to be working on a budget. Table 2-1 offers a glance at what you *could* put in your HDTV environment to make it really boom (in a good way).

Table 2-1 Home-Theater Budget Guide

Role	*Device*	*Price Expectations*
Audio sources	Tape cassette player*	$100 to $250
	CD player/recorder*	$60 to $600+

(continued)

Table 2-1 *(continued)*

Role	***Device***	***Price Expectations***
	Turntable*	$100 to $5,000+ (really!)
	AM/FM tuner*	$200 to $1,000
	Satellite radio tuner	$75 to $300
Video sources	DVD player	$50 to $1,200+
	High-definition DVD player (Blu-Ray or HD DVD)	$500+
	VCR*	$50 to $500
	Personal video recorder*	$50 to $700
	Satellite TV system*	$100 to $800
Computer/gaming	Gaming console*	$100 to $400
	Home theater PC*	$1,000+
A/V system**	All-in-one systems	$200 to $3,000+
	A/V receiver	$200 to $4,000
	Controller/decoder	$800 to $5,000+
	Power amplifier	$500+
Speakers	Center, left, right, and surround speakers	$150 to $10,000+
	Additional surround-sound speakers*	$100 to $5,000+
	Subwoofer speakers	$150 to $5,000+
Video display***	27- to 34-inch direct-view tube TV	$200 to $2,000
	Up to 65-inch rear-projection TV	$1,200 to $6,000
	Up to 120-inch front-projection TV	$1,000 to $15,000+
	32- to 60-inch plasma or LCD flat-panel TV	$1,000 to $15,000+

Role	Device	Price Expectations
Portables	Portable MP3 player*	$50 to $350
	Portable video player*	$100 to $500
Car System	Car PC*	$800 to $2,000
Accessories	Speaker and A/V interconnection cables	$50 to $1,000+
	Surge suppressor/power conditioner	$20 to $1,500
	Home media server*	$1,000+
	Internet media access devices*	$150+

** Optional*

*** You don't need all these parts, just an all-in-one system, an A/V receiver, or a controller/decoder and power amplifier combo.*

**** Only need one of these displays.*

Okay, you don't *need* all the gear in Table 2-1. You might be content with just a fabulous HDTV. Or you might be moving into a new house and want to put in a lot of the gear. Your choice.

Table 2-2 gives you a sense of what you might expect to get if you're outfitting an entire room's worth of gear for different budgets.

Table 2-2 HDTV Theater: Bang for the Buck

Price Expectations	Role
$500 to $1,000	This is the entry level for HDTV. For this price, you should be able to buy a CRT (tube TV) HDTV. You have a few choices here: You could spend the entire $1,000 and get a relatively big (32-inch) LCD HDTV, go even bigger with a 52-inch CRT-based projection TV, or go with a smaller one (26-inch or 27-inch) and add in an inexpensive Home Theater in a Box system with DVD player and surround sound.

(continued)

Table 2-2 *(continued)*

Price Expectations	***Role***
$1,000 to $2,500	Here you can begin to move up to bigger and better HDTVs. For a little more than $1,000, you can pick up a nice widescreen 42-inch microdisplay rear-projection HDTV. For a bit more, you can start getting into 50-inch and larger microprojection rear-projection HDTVs (see Chapter 22), or you could look into the plasma or LCD flat panels ranging up to a 50-inch HDTV (at the higher end of this range). In this price range, you can also begin to add fancier home theater audio solutions with larger speakers and more amplifier power.
$2,500 to $5,000	This price range is where things start to get fancy! At the bottom of this price range, you can start getting into 65-inch microdisplay rear-projection TVs, like DLP HDTVs. Near the top of the range, you can consider really big plasma TVs (50- to 60-inch models) and the largest LCD TVs (around 50-inch), and you also begin to get into front-projection systems with *huge* pictures. You can also expect that most microdisplay projection or LCD units in this price range are capable of displaying the full resolution of 1080p HDTV programming. On the audio side, you can begin to consider high-end receivers and speakers.
$5,000+	Between $5,000 and $10,000, you can build a truly top-of-the-line HDTV home theater, with a front projector, the biggest microprojector 1080p rear-projection systems, or the largest LCD or plasma flat panels. Above $10,000, the sky is truly the limit: Imagine high-end "separates" — individual amplifiers, preamplifiers, audio processors, and so on — audio equipment, the biggest and best plasma TVs or front projectors, and all the trimmings (including home theater seating, specialized remote control systems, and even motorized drapes and movie-theater popcorn machines).

One thing is for sure: Pricing is changing all the time. Two years ago, a lot of the gear mentioned in Tables 2-1 and 2-2 actually cost twice as much as it does currently. As we go to print, Sam's Club (`www.samsclub.com`) is selling a 42-inch *1080p* LCD for under $2,000. Also, a 103-inch plasma is available for a mere $50,000. (Wow!)

Go online to places like `www.pricegrabber.com` to get a sense of how much the prices have changed since we wrote this book. Use that benchmark to mentally adjust pricing throughout the book.

In deciding how much to spend overall, we can only give you this advice: Your home entertainment system is probably one of the most-used parts of your home. It helps define your family, social life, business relationships, and so on. We make a substantial investment to our entertainment systems because they get the most use.

Finding the Right Size

It's an outright crime that movie theaters can sell tickets for those rows up at the front of the cinema, where your head has to constantly move back and forth (like at a tennis match) to capture all the action. Likewise, the seats far away at the back are just as criminal.

So apply that principle to your home HDTV viewing area. You can be too close to the image (or have the image too large), or you can be too far away (or have the image too small). You definitely know you're too close if you can see the individual pixels on the screen. What you need is Baby Bear's "just right" size.

In general, experts determine the optimal size for your HDTV set by dividing the distance you are going to sit from the TV set by 2.5. (Don't ask us where they got that number, we haven't a clue — we bet trial and error. Actually, there's a lot of science regarding such technically arcane items as the size of pixels and the average person's visual acuity. You really don't want to know!) The really important thing to keep in mind is that you want to sit as close as possible to get a big-screen, immersive experience like you get at the movies, without being so close that you can make out individual pixels (picture elements) on-screen. If you sit too close, even the best HDTV with the highest quality HDTV signal feeding into it stops looking like a smooth and fluid picture and instead looks like a bunch of dots. If you sit too far, however, you're not really getting the full impact of HDTV, because your eyes just can't tell the difference between an HDTV and non-HDTV signal at great differences. It's all a balancing act.

Because HDTV pictures are so much more detailed (and because the individual picture elements for a given screen size are smaller — there's more of them in the same screen area), you can comfortably sit closer to an HDTV than you can to a standard definition set.

Table 2-3 gives you our recommendations for common screen sizes, based on your distance from the screen. (Table 2-3 is for widescreen — 16:9 — HDTVs. If you're buying a non-HDTV with the older and narrower 4:3 screen, you can actually sit farther from the screen for the same diagonal size.)

Table 2-3 Viewing Distance and Screen Size

Viewing Distance	*Recommended Display Size*
5 feet, 7.5 inches	27 inches
6 feet, 9 inches	30 inches
8 feet	35 inches
9 feet	40 inches
9 feet, 9 inches	42 inches
10 feet	45 inches
10 feet, 5 inches	50 inches
12 feet, 6 inches	55 inches
13 feet, 9 inches	60 inches
15 feet	65 inches

Matching Your HD Needs

You need to match how you intend to use the HDTV with the available technologies. Although you *can* use any HDTV type for just about any TV-viewing scenario you can think of, certain types of HDTV are better suited for particular uses. This is basically a technical issue: Different types of HDTVs use different underlying technologies to create their pictures, which often match up better with some uses than with others.

For instance, if you pack a lot of friends into a wide room to watch movies a lot, you might want to consider a plasma or direct-view CRT (a tube HDTV) display, instead of a CRT rear-projection TV or LCD flat panel. That's because plasma and direct-view CRTs have the best viewing angle (viewing from the sides). Table 2-4 summarizes other common examples driven by usage.

Table 2-4 Finding an HDTV for Your Needs

Use	*Best HDTV*
Home theater	Any projection or flat panel
Sports	DLP projection or plasma
News and stock ticker	CRT, DLP projection, or LCD flat panel
Portable	LCD or DLP front projection
Gaming	LCD, LCoS, or DLP projection, or LCD flat panel
Internet/PC	LCD, LCoS, or DLP projection, or LCD flat panel
Bedroom TV	LCD flat panel or CRT
Bang for the buck	CRT rear projection
Close to plasma for less cash	DLP, LCoS, or LCD rear projection
Showing off!	Plasma (bigger the better)

Connecting the Other Gizmos

Your HDTV doesn't live in a vacuum. An essential step when choosing an HDTV is to find a model that works within the confines of your home. That means making sure your HDTV works with

- Your chosen source of HDTV signals
- Your existing analog (NTSC) TV signal source
- Your existing video-source devices, such as DVD player, VCR, or home theater
- The *new* gear you plan to get with your leftover money (yeah right) to supplement your HDTV-powered home theater

A lot of the following info is technical stuff, which we cover briefly in this chapter. We then refer you to appropriate chapters later in the book with more detailed discussions.

Accessing your HDTV channels

To get the most out of your HDTV, you need to be able to receive HDTV channels. What you need in order to make this work depends upon what kind of HDTV you have (or are buying):

- **HDTV:** A *true* HDTV contains a built-in ATSC tuner, which can receive over-the-air (OTA) HDTV broadcasts (see Chapter 1 for details on ATSC).

  The FCC (Federal Communications Commission) has mandated ATSC tuners in all TVs, phasing it in over time, beginning with big-screen (36-inch and bigger) TVs in 2005. So eventually all new TVs will have ATSC tuners — even non-HDTVs — because the ATSC specification includes standard-definition digital TV. All TVs must have an ATSC tuner by March 1, 2007.

- **HDTV-ready TVs:** These TVs can produce HDTV images on-screen, but they don't have an internal ATSC tuner. You need some sort of external tuner to pick up HDTV broadcasts.

  We use the term *HDTV* as shorthand for both HDTVs and HDTV-ready TVs throughout this chapter — and the entire book. But when you're shopping, keep in mind that not all HDTVs have built-in tuners.

So what do you need? Well, for openers . . .

- **If you want to watch HDTV from cable or satellite sources,** you need the appropriate *cable box* or *satellite receiver* connected to your HDTV. (We talk about cable set-top boxes in Chapter 8 and satellite receivers in Chapter 9.) What's more, if you're addicted to having digital video recording a la TiVo on your receiver, you need to make sure that your service provider has an HDTV version of the DVR for your use.

  Some *DCR* (digital-cable-ready) HDTVs on the market let you watch HDTV over your cable system without a cable box. We talk more about these TVs (and the *CableCARD* that makes them work) in Chapter 8.

- **If you're going to watch OTA HDTV channels,** you need two things:
  - An HDTV antenna to receive the broadcast signals
  - An HDTV tuner (either built into your HDTV or a separate tuner) to tune into the HDTV channels

  We discuss OTA HDTV antennas and tuners in Chapter 7.

Sometimes you might want to mix and match these systems. For example, if you use satellite, you might still use an OTA HDTV antenna to pick up your local HDTV channels.

Unsure whether you want cable, satellite, or local broadcasts for HDTV channels? Chapter 6 helps you make the right choice.

Getting your analog channels

Just because you've bought an HDTV doesn't mean you can *only* watch HDTV stations with it. That would be too frustrating, given that plenty of stations still don't yet broadcast in HDTV. HDTVs are *backward-compatible* with NTSC (the old analog TV system): You can watch the analog channels and also preserve your investment in NTSC source devices such as DVD players, VCRs, and laser discs.

You can get an HDTV *set* now and then get the HDTV *channels* later. In fact, because of a device called a *scaler* (discussed in Chapter 16), you might find that your HDTV makes your non-HDTV sources look better than ever. Most HDTVs have a scaler built in, and you can also buy external scalers that are even more powerful.

Going native

Most HDTVs have one level of resolution (or occasionally, a couple of them) considered *native* to the TV. That means the HDTV is designed to display images at its specified resolution(s); any signals of a different resolution must be converted (or *scaled*) to the TV's native resolution. (Check out Chapter 1 if you're not sure what we mean by *resolution*.)

For example, many DLP and LCD HDTVs have a *native resolution* of 1,280 x 720 pixels (720p), based upon the DLP chip inside these TVs. Your HDTV probably looks its best when it is fed sources that match the native resolution, but it still looks great when the image is scaled to match the native resolution.

More important than native resolution is knowing your HDTV's *supported input resolutions* — the resolutions that the TV can actually scale to native. Some HDTVs won't do the scaling that converts between resolutions. This can be important to know when you're connecting an HDTV and an external ATSC tuner (or cable or satellite system). If 1080i is *not* a resolution that your HDTV supports — and that's all your tuner puts out — you won't get a picture. This is a rare situation, but it can happen.

Most HDTVs — whether or not they contain an ATSC tuner — contain an NTSC analog TV tuner. That way you can plug in an antenna feed and pick up all your local NTSC broadcasts. In the majority of cases, you can also tune in to *analog* cable broadcasts with this NTSC tuner.

Unless you have a DCR HDTV (discussed in the previous section), you need a set-top box to pick up *digital* cable broadcasts.

A few HDTVs — mainly flat-panel plasma and LCD HDTVs — contain no TV tuner at all. Not an ATSC tuner, not an NTSC tuner, nada! If you have one of these and you're not using a cable box or satellite receiver, you can use the NTSC tuner built into your VCR to pick up OTA NTSC broadcasts or analog cable.

If you have an external ATSC tuner for OTA HDTV, it probably also has an NTSC tuner.

Working with your other sources

Chances are very good that you're connecting more than just HDTV and analog TV broadcasts to your new HDTV. You probably want to (we're guessing) watch DVDs and videotapes, play video games, and so on.

Here are the two bits of good news:

- All HDTVs are compatible with the NTSC signals that these devices put out.
- Most HDTVs include plenty of inputs on the back (or side, or front) of the HDTV set to accommodate these devices.

Inputs galore! You really don't have to worry about "does my 1982 Betamax work with my 2007 model HDTV?" It does — as long as that beautiful old Betamax itself is still operating (and a whole underground world of Betamax enthusiasts can keep you up and running!).

The only real question is how you get all these inputs hooked up and connected to your HDTV. In Chapter 3, we discuss each of the input types themselves in more detail. Beyond understanding the types of inputs, this problem has a purely *quantitative* angle.

In effect, you need to count the devices you have (or anticipate getting) and group them together by the type of inputs they use. Then compare these numbers to the number of inputs on your HDTV. Here's a basic list:

- **Digital inputs:** You likely have one or two digital inputs (DVI or HDMI) on your HDTV. Your HDTV tuner/cable box/satellite receiver or DVD player may use these inputs.

  HDMI has pretty much replaced DVI on most new HDTVs. Though the two kinds of connectors are compatible with each other (you can use an adapter to connect HDMI to DVI and vice versa), you should choose HDMI if you have a choice.

- **Component-video inputs:** You probably have a couple of component-video inputs on your HDTV. Your HDTV tuner, DVD player, and game console (Xbox or PlayStation) can use these.
- **S-video inputs:** You probably have a bunch of S-video inputs (but you need them) on your HDTV. Your DVD player, VCR, game console, digital cable box, satellite receiver, camcorder, and even PC (yes, PC!) are just a few of the devices that can use this connection.
- **Composite video inputs:** You also have a bunch of composite video inputs. Everything we mention here can use this connection method as well.

There's also a *qualitative* angle at play here. In Chapter 3, we explain in detail, but in a nutshell, the connections listed are shown in order of rank. If you run out of inputs of a certain type (like component video) and have to use the next one down the list, you lose a bit of video quality. Therefore, it can be important, if you have a lot of gear, to choose an HDTV with *more* digital, component video, or S-video connections, if at all possible.

You can get by with fewer inputs on the HDTV if you use a home-theater receiver (see Chapter 18) that provides high-definition *video-switching* functionality. Basically, you can route everything into the back of your receiver and then use just a couple of cables to connect the receiver to your HDTV. This is a great way to avoid running out of the proper kind of inputs on your HDTV.

Which Features Matter?

A bunch of features are listed in the description of every HDTV on the market. Some are important; others are just bells and whistles that we don't think make a real difference. (We support diversity, however. If you disagree and think a certain feature is very important, by all means make that part of your buying decision! This is a very subjective area.)

Here's what we pay attention to:

- **Picture adjustments:** All HDTVs give you some degree of control over the picture settings. We like HDTVs that let you
    - Set the picture quality differently for each input on the back of the TV, so you can adjust the picture individually for the HDTV tuner, the DVD player, and so on.
    - Save multiple different picture settings in memory (like one for daytime and one for night).
- **Comb filter:** The comb filter is an internal circuit in your TV that separates the brightness and color information in an NTSC signal before it's displayed on your screen. Look for an HDTV with a *3D* or (even better) a *digital* (also called *3D Y/C*) comb filter.
- **3:2 pulldown correction:** As we discuss in detail in Chapter 21, HDTVs display their picture at a particular *frame rate* (the measure of how often the screen is refreshed every second). Typically, this is 30 or 60 frames per second (fps). Movies shot on film (in other words, most movies) are recorded at 24 fps. The process of converting film to video (called the *telecine* or *3:2 pulldown* process) converts these 24 frames to 30 frames in the same second by mixing together some of the frames, which can cause some visible blurriness or jagged lines on your screen. A 3:2 pulldown removal fixes these mixed up frames and reconstructs the original films at 24 fps. You definitely want this feature. Many DVD players perform this action on your DVD sourced video, but it's nice to have the HDTV doing it for other sources like cable TV.
- **Discrete remote commands:** Many folks who buy an HDTV integrate it into a home-theater system with a bunch of source devices, a home-theater surround-sound receiver, and more — and then they use a universal remote control to control everything without having a stack of remotes on the coffee table (see Chapter 26). These remotes often support *macros,* which are simple "one button" actions (like "play a DVD") that send many remote signals to your home-theater devices at once. When you start getting into macros and universal remotes, an HDTV that uses *discrete* remote codes (one remote signal for each action, as opposed to a single signal that cycles through a number of actions) makes it considerably easier to program and operate your system.
- **Front-panel inputs:** Do you have a camcorder (a MiniDV model, not an HDTV camcorder) or a game console that the kids are always carrying around the house? You want some front-panel inputs to connect them to so you don't have to climb behind the TV.

Look for front-panel inputs that include S-video for better picture quality.

- **Built-in speakers:** We're *huge* proponents of connecting your HDTV system to a full-up, external surround-sound audio system. We're also fans of low-impact, easy-to-use systems. So when we want to just watch the news or turn on that TiVo recording of *Sesame Street* for the kids, we prefer to use the speakers built into our HDTVs. We mention this because some HDTVs (mainly plasmas and LCD flat panels) don't come with speakers — you have to fire up the full surround-sound system for everything you watch. Also, if you plan on using a swing mount to attach your TV to a wall, you have to swing mount your speakers too if they aren't part of the display unit.
- **Surround-sound decoder:** Although you need six or more speakers and related amplification systems to get true surround sound (see Chapter 18), you can get improved sound quality for the sound system built into your HDTV if it includes a *simulated* surround-sound decoder, which can create a richer sound from your HDTV's speakers.

Service matters!

When you're buying an HDTV set, you're spending a relatively serious chunk of change. Heck, even if you're loaded, spending $10,000 or more on a front-projector system starts to move out of the pocket-change realm pretty darned fast.

So, getting your money's worth is important. That doesn't just mean getting the lowest price for the particular TV you've chosen (though that's an admirable goal — one we can definitely get behind); it also means getting decent service to boot.

Now, everyone has their definitions of what good service is, but here's what we look for:

- **Delivery services:** Lots of places (both online and local brick-and-mortar stores) are offering *white glove* delivery services. This means someone delivers that huge big-screen TV that wouldn't fit in the back of your Mini in a million years; but they don't just drop it off on the front stoop and boogie on out of there. Instead, they deliver the unit to the room you want it in, get it out of the box, and even take all the packaging materials with them (no doubt, to be humanely recycled). Importantly, if they drop it along the way, *they* dropped it and have to replace it. That's what we want!
- **Warranty:** We've never been huge fans of the extended warranty services offered by many consumer-electronics stores. After all, what's the point of paying $30 for a warranty on a $35 cordless phone? But check out the warranty that comes with your HDTV closely — it's a major investment after all. And consider an extended warranty if it's not too expensive. With some types of HDTVs (such as DLP and LCD projection systems), you might pay back the warranty when your bulb needs replacement after a few years of heavy usage.

Chapter 3

Cables and Connections

In This Chapter

- Dealing with digital video cables
- Sorting through analog video cables
- Coping with copy protection
- Plugging into audio

As you pull your shiny new HDTV out of its (probably very large) box and consider connecting it to your TV source, your DVD player, your home-theater receiver, and all the other stuff in your family room or media room, you might have a moment of panic (or at least a small shiver of fear) when you consider how many different choices you have for cables and connectors.

Never fear, *HDTV For Dummies* is here to help you. Cables are actually pretty easy when you think about them in terms of a hierarchy. Some cable/connection types are (almost always) simply *better* than others because they give you a better picture or clearer audio. When you know this hierarchy, you can quickly examine your connection options for any piece of equipment attached to your HDTV, choose your best option, and astound your friends.

In this chapter, we explain analog and digital cable options for audio and video, and we cover their positions within this hierarchy. We also explain how copy-protection systems might affect your options.

Video Connections

Look at the back of any HDTV (or any DVD player or home-theater receiver) and you see what scares many folks away from jumping in to hook up their own HDTVs and audio/video (A/V) equipment. There are just so darned many choices back there — who could possibly know which connector to use?

Well, *we* know.

High-definition video

There's often a significant difference in the functionality and video quality of connections. Only a few cables can handle HDTV signals.

Many DVD players have high-definition *connectors.* But unless you're specifically using one of the new HD DVD or Blu-ray high-definition players, you're not getting true HDTV over this connection. Instead, a circuit called a *scaler* (see Chapter 16) converts the standard DVD picture to work a little better on an HDTV screen. It's a worthwhile improvement, but it's *not* real HDTV. These *up-converting* DVD players can have a great picture, but they don't look as good on your screen as the HD DVD or Blu-ray player does. We talk about all of these different types of disc players in Chapter 11.

Digital connections

Digital video connections (such as DVI, HDMI, and FireWire) are the best choice for often-used HDTV video connections, such as the link to your HDTV from a satellite or cable receiver.

Not all HDTV devices use the same digital connections, but it's usually worth the trouble to use digital connections when you can, even if you need an adapter for *different* digital connections. As the HDMI connection replaces DVI, you might find yourself in a situation where you need to use an inexpensive (under $20) adapter. For example, if your HDTV has an HDMI connector and your HDTV set-top box or tuner uses DVI, you can connect your devices through one of these converters.

The first rule of HDTV connections: Use a *digital* video connection to connect from a source device to your HDTV, if you can. Digital connections almost always provide the best picture.

Theoretically, HDMI connections can offer the best picture quality of any of the digital connections available on HDTVs, simply because these cables have so much *bandwidth* (the amount of data they can carry) that they can offer *uncompressed* (the full HDTV signal without any potentially signal-degrading compression applied) HDTV signal transmission. In the real world, however, any of these digital connections offers an exceptionally clear and sharp picture, and HDTV signals are almost always compressed for transmission or storage anyway.

On some pieces of gear (typically HDTV cable set-top boxes, which we discuss in Chapter 8), some of the digital connections might be disabled — the connections are physically present, but the software within the device that lets them work is turned off. So, for example, if an HDTV set-top box from your cable company has a FireWire port, you probably can't use that FireWire port to connect a D-VHS recorder. Check with your cable company before spending money on cables and equipment that use these ports!

HDMI

The *HDMI* system (or High-Definition Multimedia Interface) is specially designed for HDTV connections. It carries both *HDTV video* and *digital surround sound.* Figure 3-1 shows the business end of an HDMI cable (the connector, in other words).

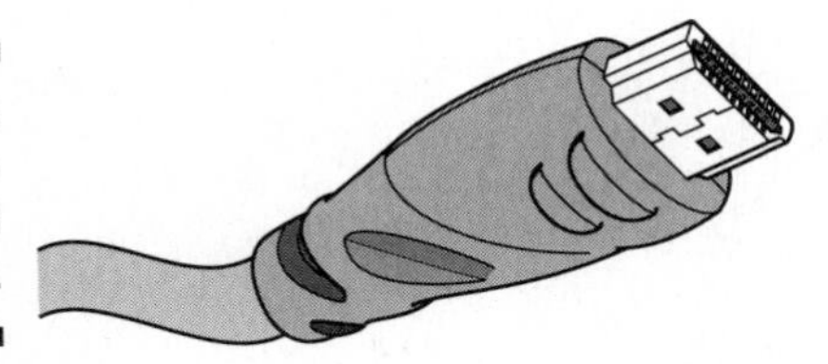

Figure 3-1: HDMI: the latest and greatest.

HDMI has a couple of advantages in an HDTV system:

- You *potentially* need only one HDMI cable to connect both HDTV video and surround-sound digital audio signals. HDMI can support up to eight channels of digital audio along with the high-definition video, all on one cable. Unfortunately, many surround-sound systems aren't ready for this "single cable" connection today, so you might need to use a separate audio cable for your surround sound. (See the sidebar titled "Can I get away with *just* HDMI?" for more information on this topic.)
- HDMI is an extremely high-bandwidth technology (5 gigabits per second). It supports all existing HDTV formats (720p and 1080i, in other words), and it has enough extra bandwidth to accommodate future HDTV formats such as 1080p. Not only does it support these formats, but it can do so without using compression — ensuring the finest picture quality you can get.

HDMI is where it's at (to quote Beck). The entire HDTV and home-theater industry is moving at breakneck speed towards incorporating HDMI in displays and other components. This is primarily because of the copy-protection systems embedded in HDMI (discussed in the next paragraph), but also because HDMI just works well and offers the highest quality of audio and video connections.

HDMI connections often require the HDCP copy-protection system for true HDTV video performance. If you're using any HDMI connections, make sure that *all* your HDTV components support HDCP — luckily, pretty much everything on the market today that has HDMI connections *does* support this. HDCP is explained in the sidebar, "No copying!"

As you shop for your HDMI cables, you might experience a bit of sticker shock: For some reason, many retailers charge an arm and a leg for HDMI cables. (We're talking $100 for a few meters of cabling!) We highly recommend that you check out some of the online HDTV sources we discuss in Chapter 10 and dig through some of the online data sources (like the AVS Forum) in Chapter 16 to find out where folks are finding good bargains on HDMI. You *can* get inexpensive HDMI cables that work really well. For example, Pat bought some cables from a vendor discussed in the AVS Forum that were about one fifth the price of the big-name brand cables in the store, and they've worked flawlessly.

HDMI cables can cover pretty significantly long distances — anything up to about 10 meters (over 30 feet) should work just fine. If you have to go farther than that (perhaps your gear is hidden in a closet on the other side of a big room from your plasma display), consult with a home-theater installation expert, who can help you figure out the best approach. After you get past that 10-meter mark, you might need to install some specialized repeater systems to ensure that your HDMI signal works correctly.

Can I get away with *just* HDMI?

Perhaps the biggest selling point of HDMI is the fact that a single cable can carry high-definition video *and* surround-sound audio. When you compare this to the analog alternative (where you can have three cables carrying the video and six or more carrying the surround-sound audio), you can easily see the appeal of HDMI.

But the truth of the matter is a bit more complex. (Isn't it always?) If you're making a simple source-to-display connection (like from a set-top box or satellite receiver to your HDTV itself), the single cable can handle your needs. But in almost all cases, you're *not* getting surround-sound audio with such a connection, because most TVs have only a pair of *stereo* audio speakers. To get surround sound, you need to mix in a home-theater receiver and speakers (see Chapters 17 and 18).

Some (but far from all) home-theater receivers have HDMI connections built in. With these receivers, you can run an HDMI cable from your HDMI-capable sources into the back of the receiver. The receiver then peels off the audio signals and sends them out to the speakers, and a second length of HDMI cabling sends the video back out from the receiver to your display. So you need two HDMI cables (or more, because you need one HDMI cable coming from each HDMI-capable source device), which is still a great reduction in the number of cables in your system.

With older home-theater receivers — and even with the majority of today's receivers — you won't find any HDMI inputs on the back. With these receivers, you need to run the HDMI cable directly from the source to your display and then run a *separate* audio cable (one of the digital audio cables we discuss later in this chapter) from the source to your receiver for surround-sound purposes.

So the bottom line is this: If you have a new HDMI-capable receiver, you *can* reach cable nirvana and use HDMI for both audio and video. If your receiver doesn't handle these connections, you need to use separate audio and video cables.

DVI

The ancestor of HDMI is the *DVI* connection. (DVI stands for digital video interconnect, and you often see the name spelled out as DVI-D — the extra D means it's for digital TV.) Figure 3-2 shows the DVI-D connector for HDTV.

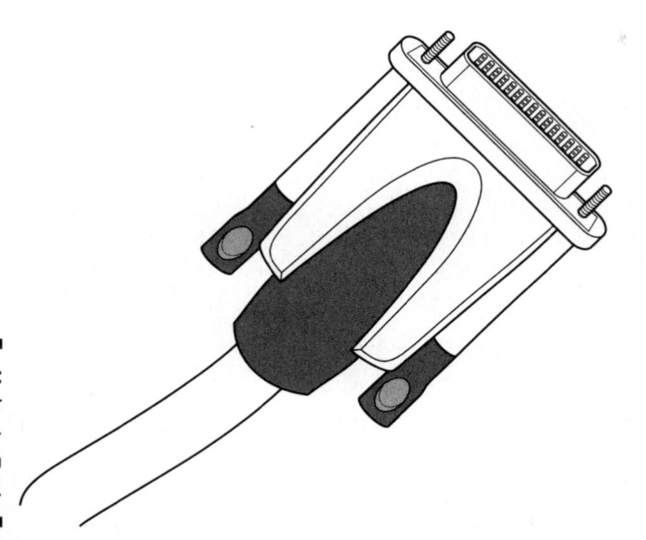

Figure 3-2: DVI-D your way to high-definition video.

If you're using DVI-D connections, watch out for two problems:

- **Not all DVI connectors work with HDTV.** Make sure you have DVI-D cables if you use DVI-D in your HDTV system.

  Computers use another type of DVI connector that has a confusingly similar name: DVD-I. The DVD-I connector has five extra pins (four pins around a central crosshair-shaped pin) on one side; these send analog video signals from computer video cards to computer monitors. You typically won't find DVI-I in home HDTV systems. (Some projection systems are also used with computers to beam PowerPoint slides onto the conference room wall.) You can use an inexpensive adapter to connect a DVD-I cable to the DVD-D receptacle on your HDTV, but you can receive only digital video signals that way, not analog.

- **DVI-D connections often require the HDCP copy-protection system for true HDTV video performance.** If just one of your HDTV components doesn't have HDCP, you might not get true HDTV performance from DVI-D connections. (We explain HDCP in the sidebar, "No copying!")

DVI-D is the only digital HDTV connection that *can't* carry audio.

FireWire

FireWire is the least-used HDTV connector. Like HDMI, it can transmit both video and audio (though FireWire doesn't have nearly the bandwidth and capabilities of HDMI). Figure 3-3 shows a FireWire connector.

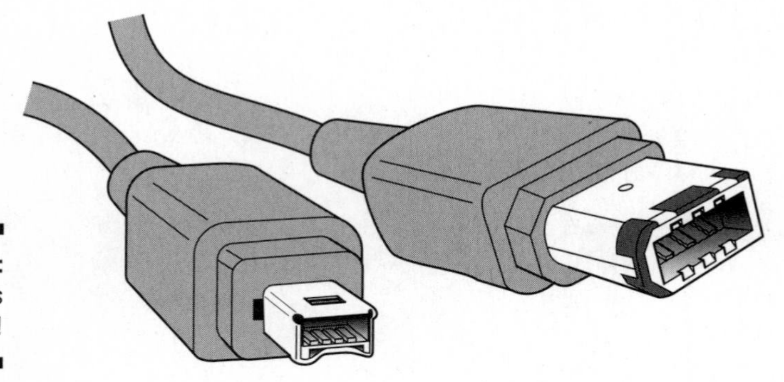

Figure 3-3: The Wire's on Fire!

FireWire is the only *two-way* connection for HDTV — the same cable can send HDTV video (and audio) to and from devices. This two-way connection is great for HDTV recording systems (for example, one cable fully connects an HDTV with a D-VHS VCR).

We use the name FireWire throughout this book, but the same system is known by a couple of other names:

- Engineers and nerds call FireWire the *IEEE 1394 Standard*. (IEEE is the Institute of Electrical and Electronic Engineers.)
- Some manufacturers use the name *i.LINK* instead of FireWire.

Consumer-electronics manufacturers usually prefer a snappy name like i.LINK or FireWire over something boring like 1394 (not much you can do to liven up that number, huh?).

We're big fans of FireWire for certain applications (like camcorders), but most HDTVs and HDTV devices use either DVI-D or HDMI connections instead of FireWire. The most common place to find FireWire is a D-VHS VCR (discussed in Chapter 13), and that's not a component that's all too common these days, because sales of D-VHS have never really taken off, and most folks use DVRs (digital video recorders, like TiVo, discussed in Chapter 12) rather than tapes these days.

FireWire isn't part of the HDCP copy-protection system. (HDCP is explained in the sidebar, "No copying!") Instead, FireWire uses its own copy-protection scheme called *5C-DTCP* (or 5 company digital television content protection), which provides similar protection of content that the big TV companies don't want you to record for yourself. The 5C system (that's the industry shorthand for it) basically acts just like HDCP, letting only 5C-authorized equipment make recordings of flagged material.

Analog component video

Component video is the only analog video-cable connection that can handle HDTV or progressive-scan DVD signals.

S-video and composite video don't carry progressive scan.

Technology

Component video is a set of *three* analog cables, as shown in Figure 3-4.

The component-video signal is divided into three (you guessed it!) components:

- *Y* is the *luminance* (brightness) signal.
- *Pb* and *Pr* each carry part of the picture's *chrominance* (color) information. Your TV uses these two chrominance signals to create the red, green, and blue colors that can be mixed together to create any color on your display. (Sometimes, Pb is labeled *B-Y,* and Pr is labeled *R-Y.*)

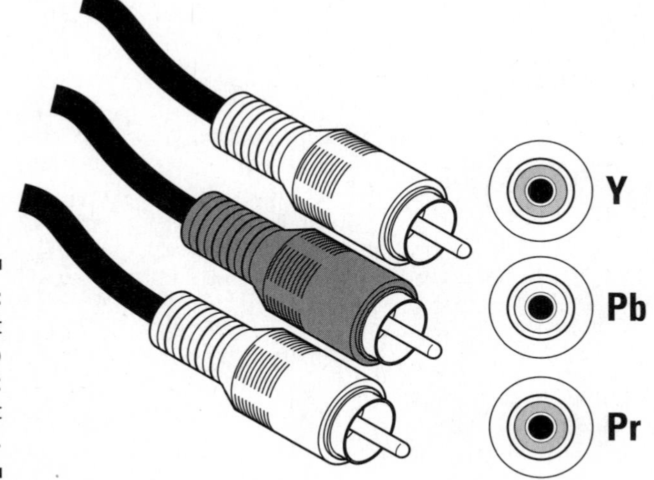

Figure 3-4: The best analog video connection: component video.

Connection

Component-video connections use three normal analog cables.

If you already have three standard *composite video* cables, you can use those cables instead of a set of "official" component-video cables. There's no functional difference between the two, though many folks find it convenient to buy component-video cables bundled together (so all the cables are neatly labeled, attached in the right order, and don't get lost).

If you're routing all of your video cables through a *home-theater receiver* (as covered in Chapter 18), check your receiver's specs before routing component-video connections through it. The receiver's *component-video bandwidth* specification should be

- At least 10 MHz for progressive-scan DVD players
- At least 30 MHz for HDTV connections

Component video might be the *only* connection that allows a true HDTV signal in your system if some, but not all, of your HDTV components use the HDCP copy-protection system. (HDCP is explained in the sidebar, "No copying!")

No copying!

Film and television studios are worried about people copying their programs and distributing them to others. HDTV (and DTV in general) is worrisome to the studios because people might be able to make "perfect" digital copies. (Non-digital copies get worse when they're recopied, so they aren't such a threat.)

These content owners have lobbied the government and manufacturers to include *copy-protection* systems in HDTV devices, such as tuners, set-top boxes, satellite receivers, and HDTVs themselves. These systems can keep you from making copies (or sometimes even *one* copy) of any HDTV program.

The most common copy-protection system is *HDCP* (high-bandwidth digital content protection). This system is in new HDMI-equipped devices and many DVI-equipped devices. HDCP *encrypts* (scrambles) the content sent between devices like tuners and TVs. This encryption is a problem if you have a DVI connection where one piece (like a set-top box) uses HDCP and another (like your HDTV) doesn't. That's because the content owners have rigged the system to *down-res* (down-resolve) HDTV programs to standard-definition unless HDCP is present at all points in the system. You could end up unable to get an HDTV signal on your HDTV!

If some but not all of your HDTV devices have HDCP, component video might be the only way to get a true HDTV signal between your devices. Component video isn't limited by HDCP, because component video isn't a *digital* signal.

Traditional video

You're probably not connecting only HDTV sources into your spanking-new HDTV. Most of today's TV content is standard definition in both *broadcast media* (over-the-air, cable, and satellite) and *prerecorded media* (DVD and VCR).

Most source devices for standard-definition video use one of the following traditional analog connectors, not the high-definition video connections mentioned earlier in this chapter.

S-video

The best traditional video connection is *S-video.* Figure 3-5 shows the S-video connector.

If you can use an S-video connection, you should (unless you can use a high-definition *digital* or *component-video* connection).

S-video cables transmit the video signal in two separate channels:

- Luminance (brightness data)
- Chrominance (color data)

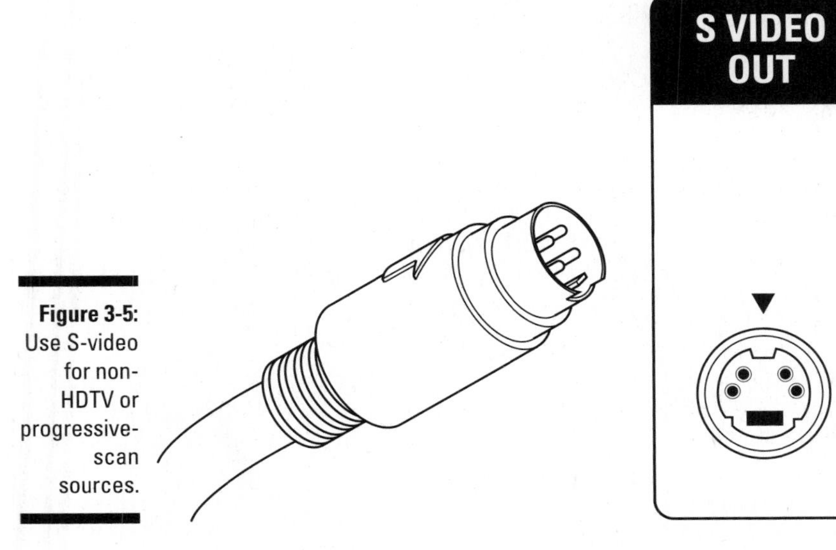

Figure 3-5: Use S-video for non-HDTV or progressive-scan sources.

Separating luminance and chrominance can deliver a better picture because it bypasses a circuit found within all but the smallest and cheapest televisions (the *comb filter*). Because an S-video cable already uses separate conductors, your TV doesn't have to separate this information with its own comb filter. Usually, this built-in separation provides a better picture on the TV.

S-video connections often are found on DVD players, game consoles (such as Xbox), and many satellite and digital cable receivers or set-top boxes. Some VCRs also have S-video connections.

S-video cables are a bit tricky to connect:

- An S-video plug has four very, *very* delicate pins and a small plastic tab. It's incredibly easy to misalign these pins and bend them (and even break off a pin). Easy does it.
- A properly aligned S-video plug easily slides in. The key is to line up the small plastic tab with the corresponding slot on the jack and *gently* push. *Don't twist!* If you're pushing hard, it's probably not aligned straight and you're probably bending pins. If the plug is hard to push in, stop and realign it.

Composite video

The oldest basic connection for separate video units is *composite video.* A composite video cable is a 75-Ohm cable with plugs called *RCA connectors* (the same connectors used by most audio cables).

Composite video cables (shown in Figure 3-6) carry the entire video signal on a single conductor. They don't carry any audio — you need separate analog or digital audio cables for that.

Composite video cables can offer a much better picture than a standard coaxial "antenna" connection. However, composite video really isn't the best choice for higher-resolution analog sources, such as S-VHS-C and Hi8 camcorders, DVD players, and videogame consoles. Use a component-video or S-video connection when you can.

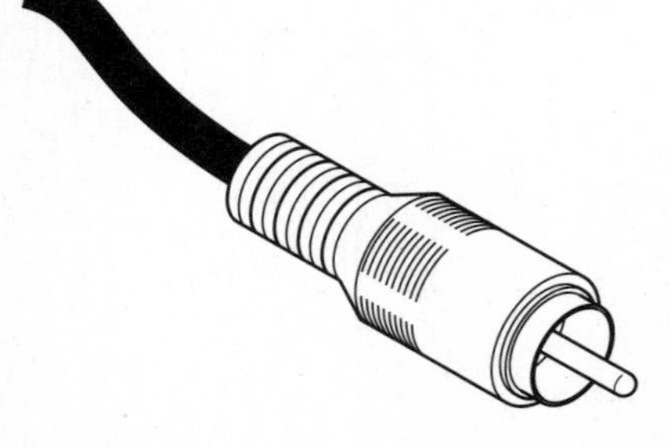
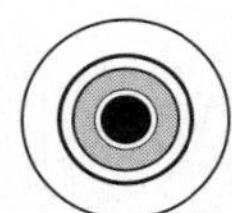

Figure 3-6: Use composite video for low-resolution video sources.

Coaxial cable

Coaxial video cable (or coax, pronounced *CO-ax,* for short) is the cable that's probably either running down from your attic antenna or coming into the side of your house from the cable company or satellite dish. It carries both video and audio signals.

A video system should use a coax connection only when you're connecting something directly to an *outside* feed (an antenna, a satellite dish, or an incoming cable line). Whenever you connect a device to your HDTV, any other video connection (even S-video or composite video) should give you a better picture than coax video.

The connectors on the ends of coax are *F connectors,* as shown in Figure 3-7. An F connector has a small pin (the *conductor*) sticking out of the middle, and a metal barrel around the outside.

Use *screw-on* F connectors if you can. Screw-on F connectors can make a much better (tighter) connection than push-on connectors.

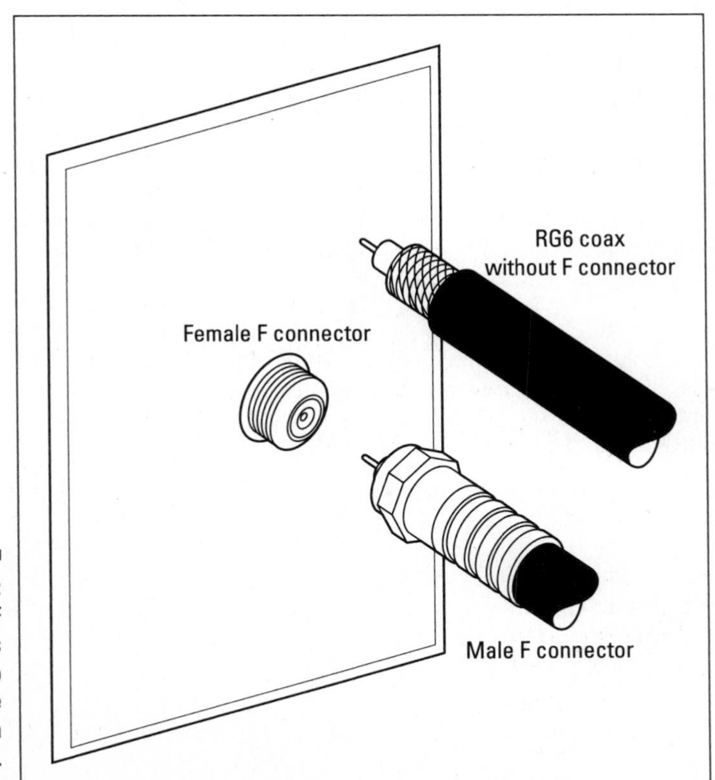

Figure 3-7: Use F connectors to hook up your cable or antenna feeds.

Audio Connections

Audio connections aren't as complicated as video. You have basically three choices, and only two deliver digital surround sound.

Every video connection for your HDTV needs separate audio connections except coaxial cable (which you should use only for *incoming* feeds from antennas, cable services, and satellite dishes), HDMI, and FireWire.

Digitizing your audio

Digital audio signals (such as those put out by DVD players and HDTV tuners) flow in a digital stream of bits from the source device to either your receiver or HDTV, which decodes this bit stream.

If you have digital connections available between two pieces of gear, use digital, not analog:

- Some home-theater gear (like DVD players or HDTV tuners) requires digital audio for the most advanced surround-sound systems (for example, Dolby Digital). If you use analog, the audio reverts to a less effective surround-sound standard.
- The longer digital audio stays digital, the better. There are two reasons for this: First, conversions from analog to digital and back again can cause signal degradation (a reduction in quality, in other words); and second, analog signals are more prone to interference, again potentially causing a reduction in quality.

The biggest advantages of digital audio are

- Near immunity to interference
- A pure digital signal all the way to the receiver

Digital audio can use either digital coaxial electrical cables or optical (Toslink) fiber-optic cables.

Optical and coaxial digital cables have very little difference in performance. The decision usually revolves around which system both your source devices and receiver happen to use.

Digital coaxial

Digital coaxial cables look basically identical to analog audio cables. They're made of similar materials, and they use the familiar RCA plug connector. The internal construction is different, however, and coaxial digital audio cables are designed to provide 75-Ohm impedance (just like video cables).

We've tried and can't hear the difference, but we thought we'd let you know that *some* HDTV enthusiasts think coaxial cables sound best.

Optical (Toslink)

Toslink optical cables are made of fiber-optic cabling (usually plastic fiber). They transmit the digital audio as pulses of laser-generated light, not as electrical signals.

Toslink cables are particularly immune to *electromagnetic interference* because they use light rather than electrical signals to carry audio. Most folks don't have an issue with this type of interference, but if your audio just doesn't seem to be working right, you can always try to swap coax for Toslink and see whether that solves your problem.

The connector on a Toslink cable is quite distinctive, as shown in Figure 3-8. It looks like nothing else in the world more than the profile of a house, with a small pin (actually the end of the fiber) sticking out of the side.

The female Toslink connector on your equipment might have a *dust cap* to keep the optical connectors clean while it isn't connected. Remove (and *save!*) this dust cap before you try to connect.

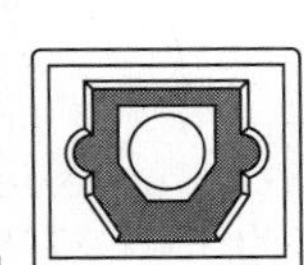
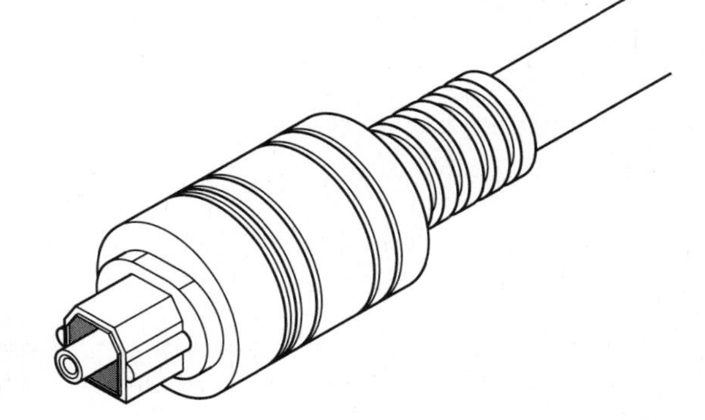

Figure 3-8: It looks like a house! Toslink audio cables.

Attaching analog audio

A couple of familiar audio connections still do a lot of the work in today's entertainment systems.

RCA cables

The basic building block of any audio connection is the tried-and-true analog audio cable (often called the *RCA cable*), as shown in Figure 3-9. You can find analog audio cables on the back of almost every source device you connect to your HDTV — ranging from HDTV tuners to 25-year-old VCRs.

Generally, you should use analog audio connections only if you can't make a *digital* audio connection.

Analog audio cables are used

- In pairs (for basic stereo sources)
- Alone (for connecting a subwoofer to a receiver)
- In sixes (for connecting either a DVD-A or SACD player to the receiver, or connecting the external surround-sound decoder on a DVD player)

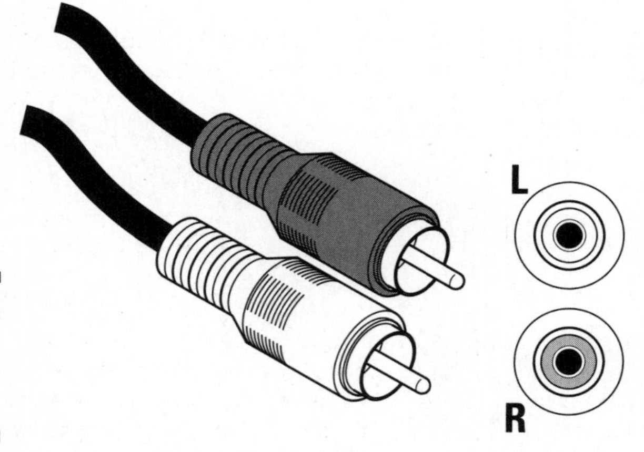

Figure 3-9: Analog audio cables.

Headphone and microphone jacks

Some video sources (such as computers and camcorders) have *headphone jacks* and *microphone jacks* instead of RCA-type audio connections.

If your gadget didn't come with adapters to connect from these jacks to an audio component that uses RCA-type connections, the nearest electronics store usually has the adapter you need.

Chapter 4

Hooking Up Your HDTV

In This Chapter

- Getting to know the backside of your HDTV
- Hooking up your most basic antenna, cable, and satellite connections
- Adding DVDs, VCRs, camcorders, and other devices
- Going digital, for all its worth

Setting up your HDTV doesn't have to be a complicated process. In fact, you should be able to test and use your HDTV with minimal setup if you want. When you start adding DVD players and home-theater systems, the connections get a little spaghetti-like.

In this chapter, we help you open up the box and get the HDTV out. (But of course you really don't believe that. Oh well. Call Pat — he'll come right on over! Kidding.) Okay, so what we *really* help you do is understand the overall layout of your HDTV system and understand where the parts fit. We start small and simple, with just an over-the-air connection, and work up to your most intense and complex HDTV home-theater operation.

Throughout this chapter, we refer to all the connections discussed in Chapter 3. You can flip there for the pros and cons of a particular cable or interface.

The exact ports, connections, options, and other attributes of your HDTV set vary by manufacturer, model, country, and so on. So we use a rather generic diagram to illustrate your HDTV set's interfaces, as shown in Figure 4-1. In practice, you might not have all of these, or they might be scattered around on the sides of your HDTV (which often happens with a plasma) or across the back (which you get with a projector). The key point is keeping track of what you're connecting — and with which cables — so don't sweat it if Figure 4-1 doesn't look exactly like your TV.

Taking a Look at the Different Connections

Unless your HDTV has a built-in DVD player or VCR, you need to make some connections to actually use it — the video signal must come from somewhere. There's probably a cluster of input and output jacks (a *jack panel*) on your HDTV. The devices you connect to your HDTV determine how you use these jacks.

We assume that you're connecting your source devices directly to your HDTV. In some cases, if you have a surround-sound audio system and a home-theater receiver in your setup, you might end up connecting some (or all) of your source devices directly to the receiver and then connecting the receiver to your HDTV. We work up to that scenario as we go through the following process of looking at inputs and connections.

Depending on the cabinet design, jacks might be on the front, at the side, or on the back of your set — or even in all of these places. Figure 4-1 shows the input and output jacks you're likely to find on the back of a CRT or projection HDTV.

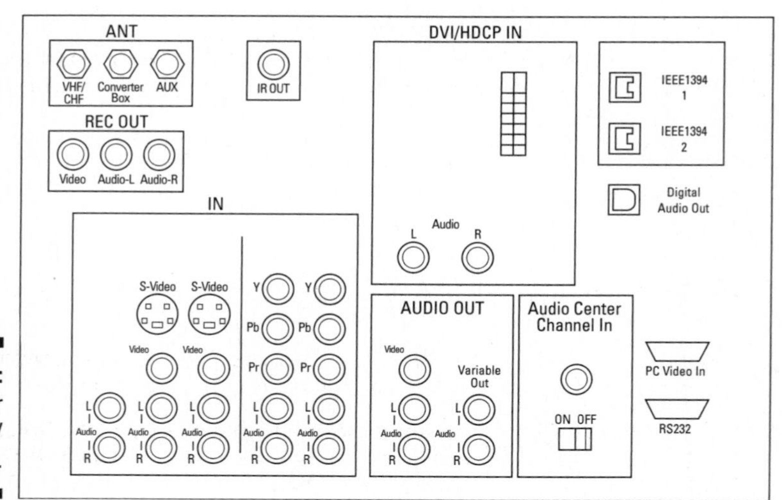

Figure 4-1: Your master set of HDTV ports.

DVI-D, composite-video, S-video, and component-video connections carry video information only. You need separate audio cables to carry the corresponding audio signals — something you won't need with an HDMI connection. Chapter 3 covers these connections.

Input jacks

Input jacks only *receive* broadcast signals and programs from your audio/video devices so you can watch them on your HDTV. You'll probably see most or all of the following audio/video input jacks on your HDTV (refer to Figure 4-1):

- **ANT-IN:** Two or more ports for
  - NTSC analog and ATSC off-air signals

    These ports work for ATSC only if you have an HDTV with a built-in HDTV (or ATSC) tuner.
  - Analog and digital cable-TV signals

    These ports work for digital cable only if you have a DCR (digital-cable-ready) HDTV with a built-in *QAM tuner* (see Chapter 8).
- **HDMI In:** Most HDTVs today have at least one, if not two, HDMI inputs, which can accept the output of an HDTV set-top box or receiver, an up-converting DVD player, a Blu-ray or an HD DVD disc player, and some other devices (such as Sony's PlayStation 3 game console). HDMI handles both video and audio over a single cable, so you don't need a separate input and cable for the audio portion of these signals.
- **DVI-D/HDCP IN:** A digital-video input, usually teamed with two R/L inputs for audio.

  DVI-D/HDCP ports can't plug into cables that are connected to a PC's (similar) DVI-I connection. Nothing will blow up, but you won't see a picture either!
- **Video In:** Typically these inputs are in sets with
  - Composite-video and S-video inputs
  - Standard audio inputs

  In most cases, these inputs are connected to composite or S-video–equipped video systems such as VCRs.

  Your HDTV might require you to tell it whether you're connecting to the composite-video or S-video jack. (Check your owner's manual.)
- **Component Video In:** Component-video plus standard-audio inputs for accepting signals from component-video systems such as DVD players, HDTV set-top boxes, and gaming consoles.
- **PC inputs:** Usually divided into
  - *PC Audio Input:* These audio jacks connect to the audio output ports on your PC.
  - *PC Video Input:* These video jacks connect to the video output port on your PC.

Dealing with the "no picture" scenario

Most HDTVs have a fairly large number on inputs on the back (and usually some on the front, too!) that can accept connections from antennas, cable TV feeds, cable set-top boxes, satellite receivers, DVD players, DVRs, and more. Having all these connectors is a good thing, but it can occasionally cause a bit of trouble.

So after you connect one of your source devices to your HDTV, you might run into a situation where you start up your HDTV and expect a picture and get . . . well . . . nothing. Don't panic. Most TVs don't automatically search amongst all the inputs you have hooked up to find the one that's active. So read the manual for your shiny new HDTV, and as a first step — before you learn anything fun, such as how to operate the picture-in-picture function or the zoom function — figure out how to select the various inputs.

Some HDTVs have *assignable inputs* so you can tell your HDTV and remote control that, for example, Input "1" on the back (which might be a component-video connection) belongs to the DVD player and should be activated when you press the DVD button on your remote, and so on. Configuring your assignable inputs (and assigning them labels that make sense to the average human being!) takes only a few minutes, and it's worth getting figured out right away. If you're connecting your video sources through an AV receiver, you often run into this situation, as well. Just use the receiver's on-screen display to have it learn which inputs correspond to which devices. Note that this applies to both video and audio inputs.

Output jacks

Output jacks *send* signals from your HDTV to your audio/video devices so you can

- Record, listen to, or distribute the programs
- Control other devices

You probably can see most or all of the following audio/video output jacks on your HDTV, as shown earlier in Figure 4-1:

- **IR Out:** An infrared port for sending IR signals to control your attached devices.
- **Audio/video outputs:** Usually an HDTV has two kinds of audio/video outputs that do a range of tasks:
    - *REC Out:* A "record out" connection for recording what you see on your HDTV to an analog VCR.
    - *A/V Out:* Regular composite-video and standard-audio outputs for connecting such devices as a VCR for editing and dubbing. Usually this output is *bridged* directly to an input; whatever is connected to the input jack goes to this output.

These audio/video outputs usually have a couple of limitations:

  - They output a *down-converted* video signal, not HDTV.
  - You can't adjust the audio volume with the TV remote.

- **Audio Only:** HDTVs with a built-in HDTV tuner (ATSC or QAM) usually have a couple of outputs for sending audio to other devices, such as amplifiers, receivers, and decoders:
  - *Digital Audio Out:* A digital audio connection (usually an optical "Toslink" connector) for connecting external Dolby Digital–enabled amplifiers, receivers, decoders, or other home-theater systems that receive optical audio.
  - *Variable Audio Out:* These are standard analog audio ports for connecting an analog amplifier with external speakers.

    Variable audio allows you to adjust the volume of your external sound system with your TV remote.

Many TVs have an on/off switch or setting within the on-screen display setup menu that governs how the onboard speakers are used. You might be able to switch your speakers so that either

- The internal speakers carry all the normal audio signals.
- The TV's audio goes directly to the A/V receiver, and either
  - The TV's speakers stay on playing back a stereo soundtrack.
  - The TV's speakers are entirely off.

Bi-directional jacks

Bi-directional jacks both send and receive data. A couple of these jacks are common on HDTVs, as shown in Figure 4-1. Two types exist:

- **FireWire/IEEE 1394:** These ports are for connecting devices for compressed video and audio signals.

  HDTVs with IEEE-1394 FireWire ports don't always operate with all the other devices that include such connections. For instance, usually you can't hook up a 1394-outfitted Mini DV camcorder to a TV. Check your manual closely about what can be connected to this port.

- **RS-232 Jack:** This is a serial-connection port that you can use with your PC for data transfer (like *firmware* upgrades) and with some automation systems.

Connecting Basic TV Sources

When you get your HDTV home from the store, we know you'll want to turn it on and fire it up. You don't have to connect *everything* right from the start. The following sections cover the most basic connections you can make with your HDTV.

Throughout this chapter, we deal with VCRs and standalone DVRs as interchangeable units — they connect to your HDTV in the same fashion. If you have both, see the section called "Two VCRs for editing."

If you want to connect your VCR or DVR right now, skip to "Connecting DVRs and VCRs," later in this chapter.

Antenna or unscrambled cable

The simplest connection you can make is connecting an antenna to your HDTV — with no worries about cable boxes, VCRs, DVDs, home-theater receivers, or anything else. You use a direct antenna connection when

- You don't need an external TV tuner to tune into NTSC or HDTV channels because your HDTV has an ATSC (for broadcast) or QAM (for cable) tuner built-in.
- You don't need a cable box or satellite receiver to unscramble channels.
- You don't connect a VCR or DVR.

This setup works only if your cable channels are unscrambled *or* if you have a CableCARD system in your HDTV to unscramble premium channels. See Chapter 8 for more on the CableCARD.

This connection is shown in Figure 4-2.

When you use an antenna to pick up HDTV or standard-definition broadcasts, it is called using an *OTA* (or over-the-air) connection.

If you're using an indoor antenna to grab either HDTV or regular, standard-definition OTA signals, make sure you keep the antenna away from the TV to avoid noise on the screen.

Cable box

Your best cable-television connection depends on how you (or whether you need to) use a cable box to access cable channels. If you don't need a cable box, follow the instructions in the preceding section.

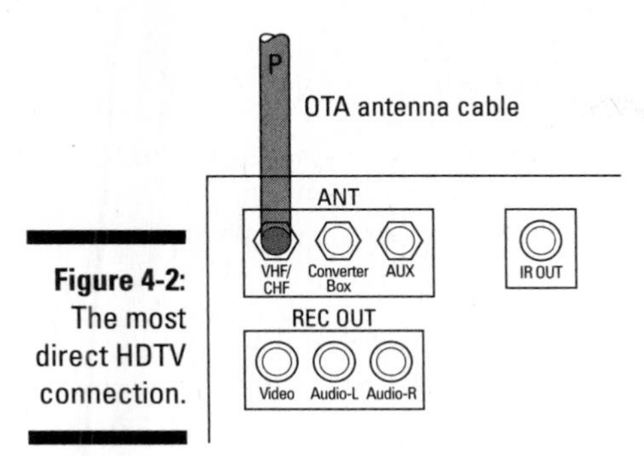

Figure 4-2: The most direct HDTV connection.

We're talking about *analog* cable boxes here. If you have HDTV service from your cable company, skip to the section titled "Digital cable or satellite connections" for information on how to connect a cable or satellite box to your HDTV.

Cable set-top box option 1

This option, shown in Figure 4-3, adds a cable set-top box to the equation. You would use this connection plan if

- Your cable company requires a cable set-top box *only* for scrambled or digital channels.
- You don't connect a VCR or DVR.

Even though most cable set-top boxes have an RF output that can connect to the ANT (antenna) input on the back of your HDTV, you're better off using the component-video connections on your cable box and HDTV, if they're available. (See Chapter 3 for more details on these connections.)

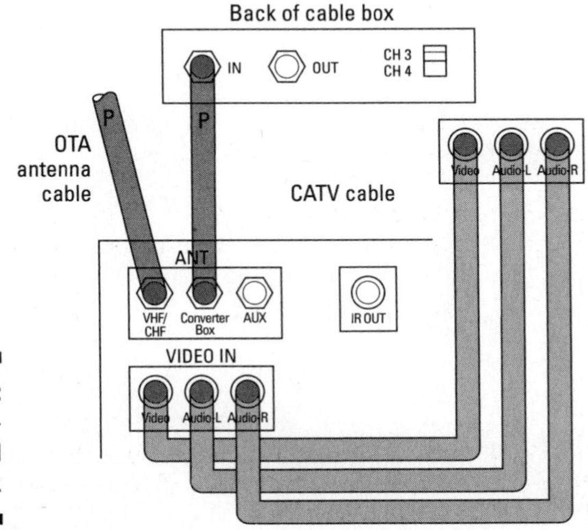

Figure 4-3: Cable set-tops and your HDTV.

Most of the time, you use the set-top box for all your cable TV viewing. This option can come in handy, however, if you have *PIP* (picture in picture) capabilities in your HDTV and you want to be able to watch the unscrambled channels in one PIP window and the scrambled ones in another.

The HDTV's internal converter eliminates the need for an external splitter, so you can switch between two options:

- Unscrambled signals come straight into the TV set.
- Scrambled signals come in through the cable set-top box.

Cable set-top box option 2

The option shown in Figure 4-4 consolidates all your signals over one link to the HDTV set. You would use this connection option if

- Your cable company scrambles all its channels, requiring you to have a cable set-top box.

  This situation limits how much you can actually use TV-set features that might allow you to view two channels at once (that is, *picture-in-picture* capability). If the cable set-top box sends one channel at a time, that's all you can watch.

- You don't connect a VCR or DVR.

  If your HDTV is DCR (digital-cable-ready), you might be able to use a *CableCARD* and skip the cable box. We discuss this choice in Chapter 8.

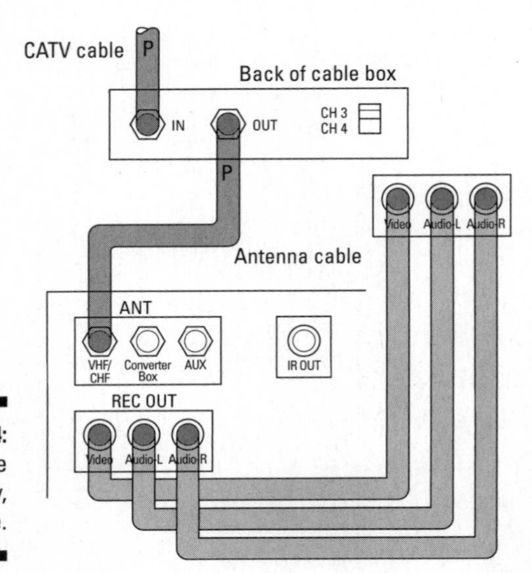

Figure 4-4: Going cable all the way, all the time.

Satellite receiver

The satellite receiver option, shown in Figure 4-5, can consolidate all your signals over one link to the HDTV set if you like. You would use this connection option if

- You have a satellite service and receiver.

  The signal from the satellite dish connects to the receiver, which in turn connects to the HDTV.

- You have an over-the-air antenna.

  The antenna cable connects to either

  - The HDTV (via one of its antenna ports)
  - The satellite receiver

- You don't connect a VCR or DVR.

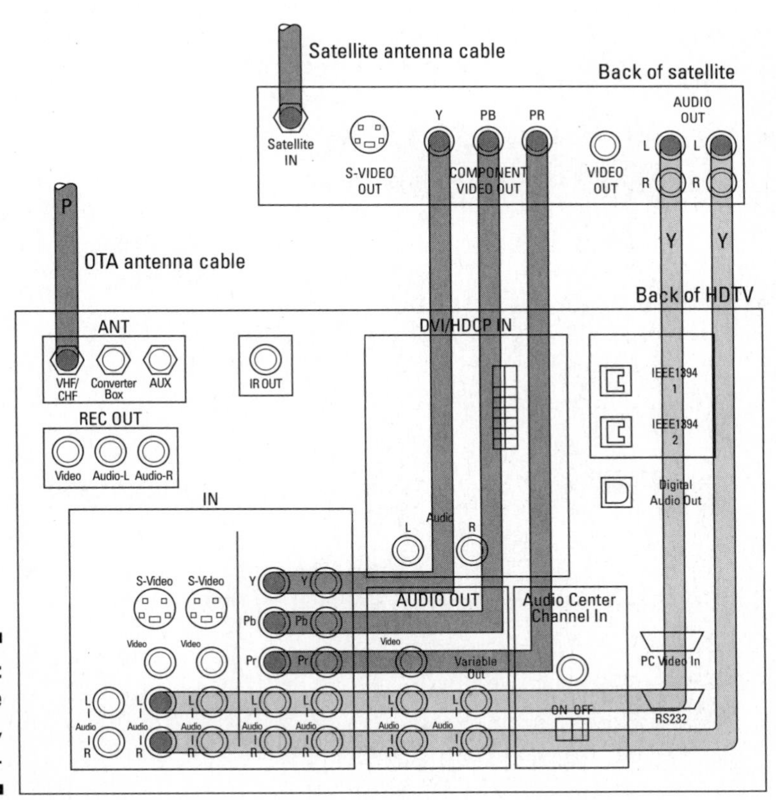

Figure 4-5: Beam me down, HDTV-style.

Most new satellite receivers have HDMI or component-video connections in addition to composite and S-video connections. HDMI is always the best of these options for this application, but component video is nearly as good, and you should definitely use it if you can't use HDMI.

Connecting DVRs and VCRs

If you have a DVR or VCR, your connection options increase. The upcoming subsections describe some potential connection situations you might encounter.

We consider the DVR and VCR to be interchangeable items — which is exactly what they are, functionally speaking. So in the following sections, we talk about DVRs, but you can use the same hookups for VCRs.

If your cable or satellite set-top box or receiver has a built-in DVR, you don't need to make any extra connections at all. All the connections are internal to the set-top box or the receiver.

DVR/VCR with antenna or unscrambled cable

The most basic DVR/VCR option (see Figure 4-6) is to run an antenna feed or cable TV connection to the VCR and then on to the HDTV. You would use this connection if either of the following is true:

- You're using an OTA antenna.
- Your cable company doesn't scramble its signals (so you don't need a set-top box).

If you have both an OTA antenna and cable TV, connect the cable TV to the UHF/VHF (also called ANT 1) jack and the antenna to the AUX (also called ANT 2) connection. Although this might seem counterintuitive, most TV features are set up to use the UHF/VHF connection as the primary default. That's why you'd want your cable connection on this port.

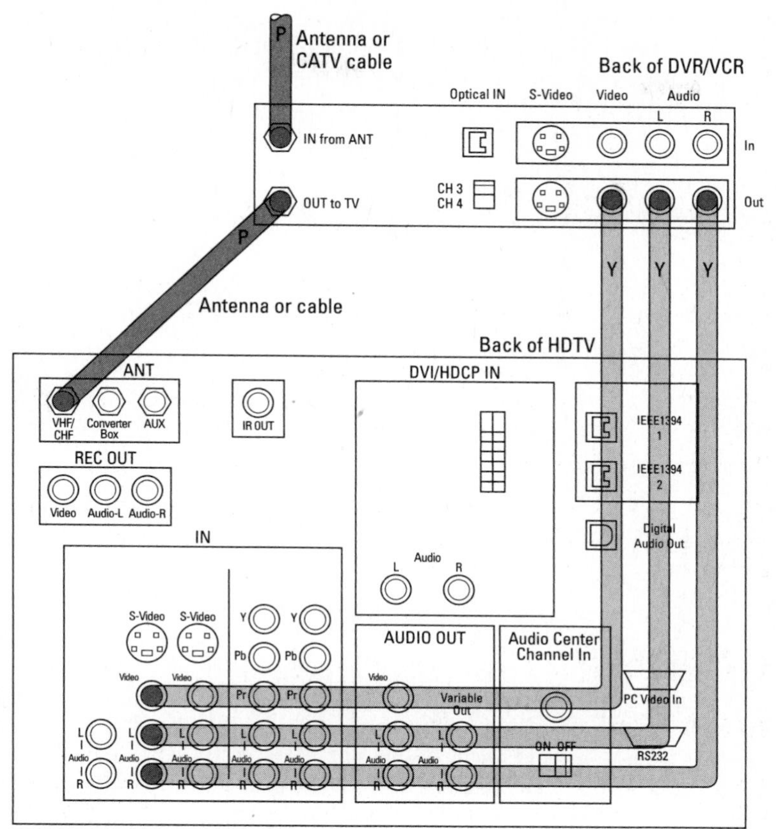

Figure 4-6: A DVR/VCR and your direct cable connection.

DVR/VCR with a cable set-top box

This DVR/VCR option (see Figure 4-7) is to run a cable TV connection to the device and then on to the HDTV. You would use this connection if

- Your cable company scrambles some of its signals but not all of them.
- You want to connect a DVR or VCR to pick up these scrambled signals.

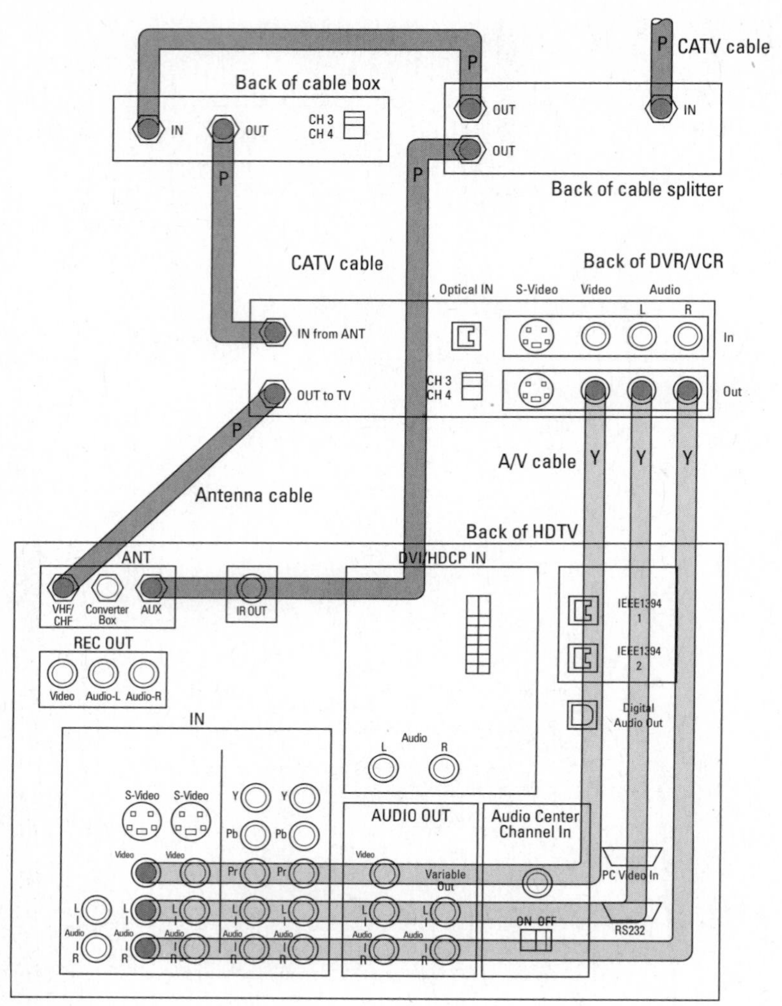

Figure 4-7: A DVR/VCR and cable set-top box in tandem.

A splitter takes the signal from the cable company and sends identical signals on two paths:

- Directly to the TV's ANT 2/AUX antenna port
- Through the cable set-top box to the TV's S-video or composite-video inputs

Using two antenna connections lets you use the TV's tuner on unscrambled channels for such features as picture-in-picture. The cable set-top box usually sends only one channel at a time, so you can't use most of the TV's multi-channel features with it.

DVR/VCR with a satellite receiver

This DVR/VCR option (see Figure 4-8) is to run a satellite TV connection to the device and then on to the HDTV. You would use this connection if

- You use a satellite company for your main signals.
- You have a DVR or VCR that is *not* internal to your satellite receiver.
- You have an over-the-air antenna.

Most new satellite receivers have component-video connections, in addition to composite-video and S-video connections. Component is always the best of these three options.

In this arrangement, you can use the DVR to view or record antenna-based channels, in addition to viewing the antenna channels through the HDTV.

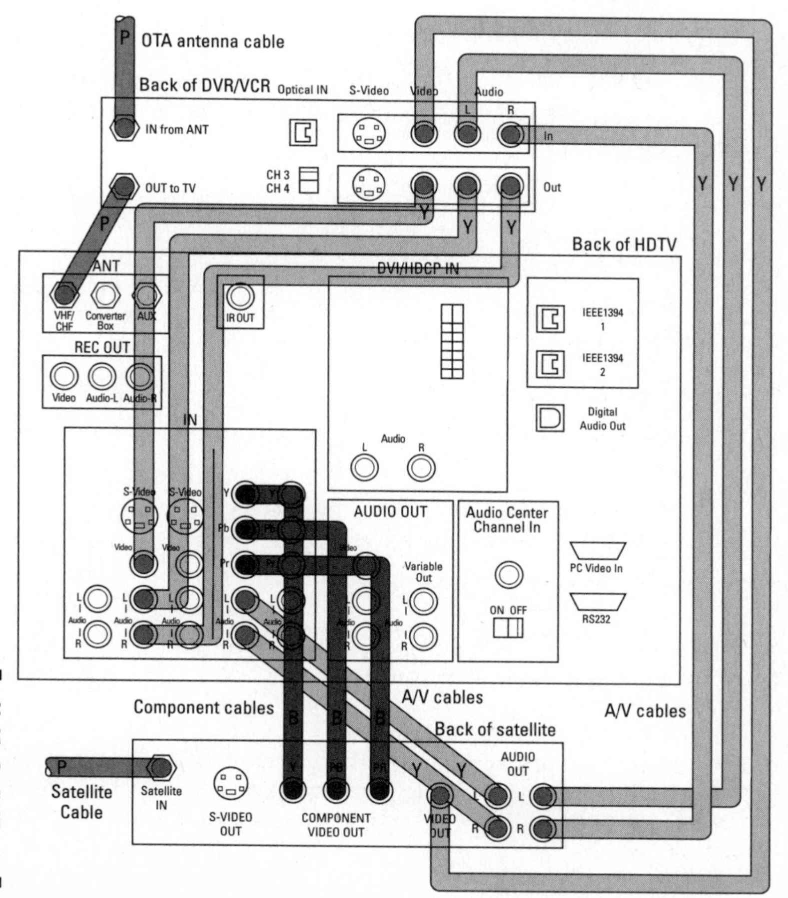

Figure 4-8: A DVR/VCR mated to your satellite receiver and HDTV.

Making Other Connections

The fun doesn't have to stop here. You can add more to your HDTV setup.

DVD player

You can add a DVD player to the cable TV service, antenna, satellite, DVR, VCR, or whatever it is you have hooked up to your HDTV quite easily. In fact, what you have connected to your HDTV doesn't really matter. The source device (such as a DVD player, video game console, or laserdisc player) connects to its own set of audio and video inputs on your HDTV.

Use the highest-quality video and audio connection available to you (see Chapter 3 for details on this). Whenever possible, use HDMI or component video and digital audio cables. Move down from there based on what's available on your HDTV and DVD player.

If you have a VCR/DVD combination unit, you probably have to make separate connections for the DVD player and the VCR, just as if they were separate devices.

Two VCRs for editing

If you do a lot of taping and editing, you might have two VCRs connected together (as in Figure 4-9). You can either

- Connect two VCRs *serially:*
  - The playing VCR's Line Out (video and audio) ports connect to the Line In ports on the recording VCR.
  - The recording VCR's Line Out ports send the signal to the HDTV's Video and Audio In ports.
- Hook up the playing VCR to the Line In ports on the HDTV and then connect the recording VCR to the Line Out jacks of the TV.

  On most TVs, the Video Out jack doesn't output a picture-in-picture frame, so you can't record that image.

If you're connecting two VCRs together for tape editing, you still can't edit a tape that has copy protection.

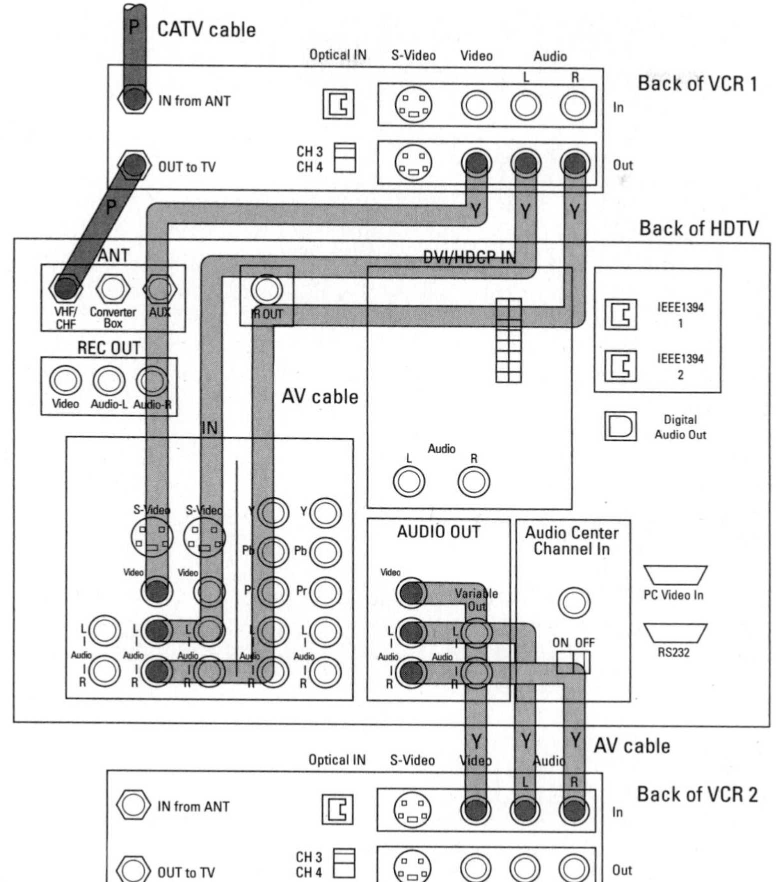

Figure 4-9: Two, two, two are better than one.

Getting your PC hooked up

If you have a Media Center PC (or a homemade HTPC — home-theater PC), you definitely want to connect it to your HDTV. How you make this connection depends upon two factors:

- What kind of video card you have in your PC
- What kind of connections are available on your HDTV

The video card determines what kinds of connections are available on the PC; the back of your HDTV determines the other end of the connection. Some HDTVs can connect directly to PCs, and have a VGA connection. If you don't have one of these connections, you want to use component-video or S-video cables to connect your video, as well as a pair of analog audio cables.

If you have either a Media Center PC or a PC with DVR software, you can treat your PC like a VCR or DVR (as described in this chapter).

Digital cable or satellite connections

If your set-top box is digital, you most likely have component-video connections — and potentially even HDMI connections (check out Figure 4-10). In either case, you have to connect audio cables separately. The rest of the connections to other devices are the same as discussed elsewhere in this chapter.

To make sure your HDMI device resets correctly, we recommend that you turn on the TV first and then power up the HDMI gear. When you turn them off, shut down the HDMI device first.

Camcorder

These days it seems everyone has a camcorder. If you have a camcorder, you can connect to your HDTV via S-video if you have it; use a composite connector if you don't have S-video. (Figure 4-11 shows what a camcorder connected via composite video looks like.)

The audio outputs of your camcorder connect to a pair of analog audio inputs on your HDTV.

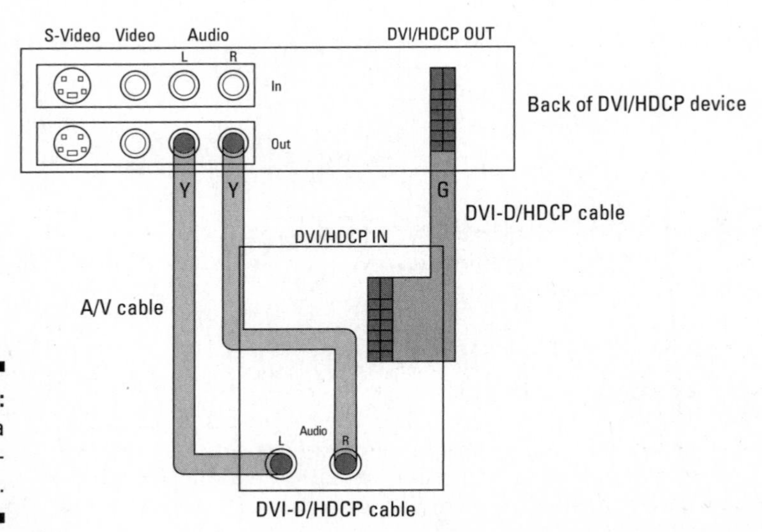

Figure 4-10: Going with a digital set-top box.

Most HDTVs have *front-panel* inputs (usually one for S-video, one for composite video, and a pair of analog audio inputs) designed especially for connecting camcorders — you don't have to bend over in an unflattering way behind the TV to make your connection.

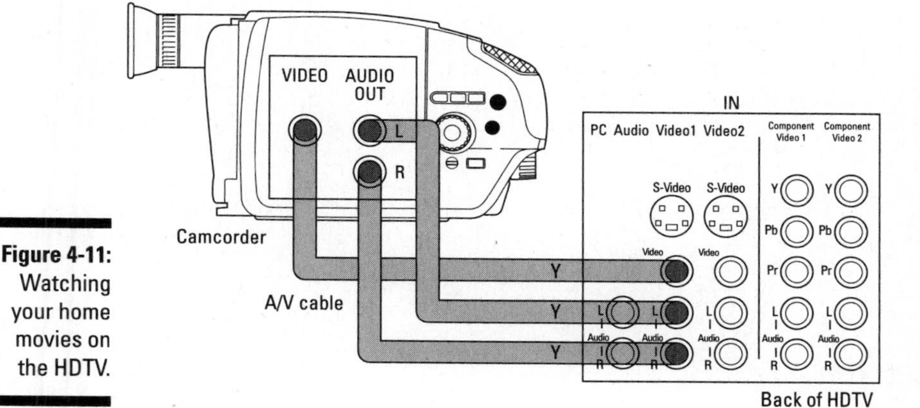

Figure 4-11: Watching your home movies on the HDTV.

Surround-sound system

You can — and we hope you do — connect your HDTV to a full-fledged home stereo system. This is shown in Figure 4-12. You would use this scenario if

- You want full surround sound from your HDTV experience.
- You have a home-theater A/V system rated for surround sound.

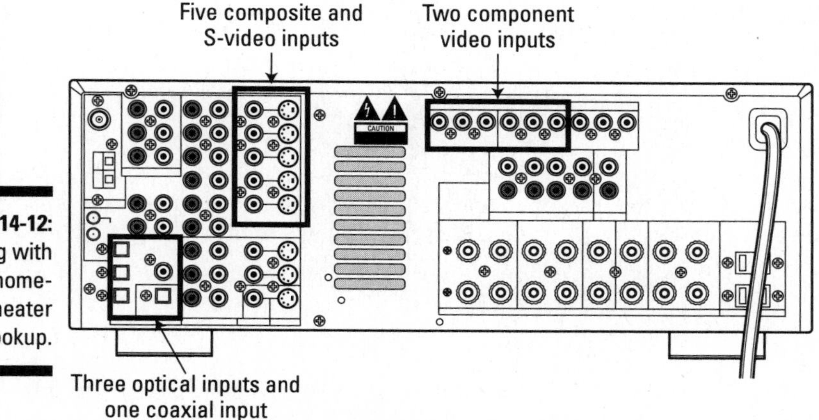

Figure 14-12: Going with the home-theater hookup.

Picture-in-picture with cable boxes

When a cable system requires you to use a special cable box (for example, to access premium, digital, or pay-per-view channels), your TV receives only one channel at a time from the box. That means you can't use some of your TV's special features — in particular, picture-in-picture. If all your cable or satellite channels require the box, you can't do much about it. (You might be able to get a cable or satellite box with these features — a *two tuner* model — if you want them.) But if your cable system has some *unscrambled* channels, you can add an extra cable input directly to your TV that lets you use all of your TV's features on unscrambled channels. You can set up your system this way with any of the cable-box connections shown in this chapter.

This arrangement requires a cable splitter and a couple of extra 75-Ohm cables. Instead of connecting the incoming cable feed directly to the cable box, you connect the incoming feed to the splitter. The extra cables connect the TV and cable box inputs to the splitter. The following figure shows this extra connection with a cable box. Note that with this connection, the TV has direct access to any unscrambled cable channels for such features as picture-in-picture.

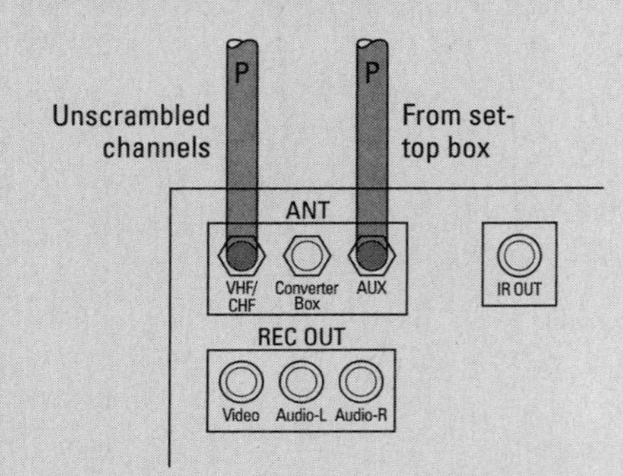

Basic channels are hooked up to an antenna input on your TV. To tune these channels using your TV's picture-in-picture functionality, just select that antenna input. You can watch *all* your channels — basic, unscrambled channels, and premium channels — through your cable box, which should be connected to your HDTV using composite or S-video cables (along with a pair of audio cables).

This is actually quite simple in design — you're taking the video and audio out from the TV and using them to drive what comes out of the stereo system. Note that if you're using your receiver as a video switch, you would instead route the video signals into the receiver *before* they go on to the HDTV.

If you plan to use a receiver as a video switch to control the signals going to your HDTV, realize that the highest-quality digital HDTV connections (HDMI, in other words) are often *not* supported by receiver video switching, so you need to make those connections directly from source to HDTV, and not use the receiver's video switching. More on this in Chapter 18.

Chapter 5

Mounting Your HDTV

In This Chapter

- Swinging with your TV in two rooms
- Finding the right mount for your circumstances
- Installing your HDTV just right
- Disguising your HDTV as your favorite museum artist's painting

It used to be that, in order to have a big-screen experience, you had to buy a massive rear-projection TV that was almost as deep as it was wide. These old-style units took up one whole end of any room and clearly were the source of many spousal arguments during football season (or any time for that matter, but many of these sets were delivered just in time for the big bowl games!).

These times have passed. You can mount thin LCD and plasma panels almost anywhere. Swivel and universal joints on mounts allow you to tilt or rotate screens with the mere touch of a finger. New standards among manufacturers regarding mount locations on the display frame have buoyed a third-party mount market that now gives you more choices than ever.

So it's time to think outside the (TV) box and consider ceiling or wall mounts, lifts that hide your TV when not in use, or any of a number of other approaches. Read on!

Planning Your HDTV Mount

Like any construction project, it pays to plan and design well upfront. Mounting a TV gives you more options about how you might use it; likewise, how you might use the television can dictate the mount you buy and install.

You want to mount a TV only once. Plan ahead, and you shouldn't have to do it a second time. Some things to think about when planning your installation:

- **Are you planning on mounting your display over a fireplace?** Fireplace installations take a lot of care because too much heat can damage your display. You can mitigate damage by buying a TV with an inside fan for cooling. (A lot of TVs don't have a fan for cooling — they use *convection,* or air flow, for cooling and have to be five inches from the wall to allow proper air movement.) If a lot of heat is coming up from the fireplace, you probably need a mantle to deflect the heat.
- **Are you planning on buying external speakers (that is, speakers that are separate from your display)?** If so and if you intend to use a swing mount, make sure you can purchase an add-on accessory to mount speakers onto the display. Think about the center channel speaker that you don't want bolted to the wall when your display has swung away. You want to avoid having the lips of the characters on your HDTV moving in one place with the speech coming from another — yech!
- **Do you have the right foundation or clearance for a mount?** If you have only wallboard in your walls, you don't want to risk mounting a $4,000 HDTV by using only lag bolts if you can help it. The heavier the TV, the more foundation support you need to make sure it doesn't pull away from the wall. You really need to be able to attach your mount to wall studs to safely support a larger monitor.

This list is just the appetizer. Table 5-1 gives you loads more ideas for your planning.

Your mount manufacturer should have some very explicit instructions and guidance on how to safely mount an HDTV. If the mount you're considering doesn't come with instructions, we suggest you shop for another mount!

If your space is constrained, one of our favorite sites has a feature for you: Check out the Crutchfield.com *TV Fit Finder,* which helps you fit a TV to your specific spot. You choose from two options ("I need a TV that will fit my stand or cabinet" or "I need a stand or cabinet that will fit my TV"), and the system helps you find those products available on the market that meet your needs. We love the Internet!

Table 5-1 Think and Plan before You Buy a Mount

Ask Yourself . . .	*You Should . . .*
Do you want to share the TV across rooms?	Use a swing arm mount so you can merely rotate the TV between rooms. (See Figure 5-1.)
Where do you want to put the display in the room?	Plan for your cabling outlets in advance.

Ask Yourself . . .	***You Should . . .***
Are you mounting the TV on a brick wall or in front of window?	Opt for a ceiling mount, as it's easier to run your cables invisibly.
How do you want to use the TV when it's installed? Will you view it lying on the floor sometimes and on the couch other times?	Buy a swing mount with a high degree of pitch so you can change the angle as you desire.
Do you want to press a button and have the display automatically appear?	Look for automated pop-up and ceiling products for this purpose. These are great for places like apartments in the city where you might have limited space.
Do you care about specific orientation of the display?	If you're into photography and want your display to go into portrait mode for vertically oriented pictures, get a mount that allows you to rotate your display into vertical position. Such mounts also allow you to view the display horizontally but store it vertically, if space is a consideration.
Is this room subject to different lighting during different times of the day?	Consider pitch wall mounts — ones that can shift a display up and down — to avoid the glare.
Is this a permanent installation or something you might change in the mid-term as newer TV display models come out?	If this mount is a temporary installation, choose a mount that can likewise be temporary if you might change to a different approach in the future. (Automated lift mounts are usually pretty permanent installations.)
Is décor or look an issue?	If you need the mount to be really flush with the wall with all your cables well hidden, use a swing arm mount because it allows you to use a recessed area in the wall. You can connect the mount to the wall, pull out the mount and connect all the cables and display, and then push the mount back against the wall. Voila! You have a fashionable flush installation. (You can't do this with a static mount; it's impossible to connect all your cables.)
Do you need to hide the display itself?	Look into artwork options that slide over the display or that add a decorative frame-like trim. You can even buy a material cover that acts like a mirror when the display is off and is transparent when the display is on.

Figure 5-1: Chief Manufacturing's Reaction PDR-U Dual-Arm swing mount covers 270 degrees.

Choosing a Mount

You run into three basic types of mounts:

- **Flat panel:** Designed for plasma and LCD displays predominantly, you find versions for ceiling, wall, and pole mounts (for floor-based mounting). Flat-panel mounts can be static, pitch, and swing mounts. Static mounts don't move. Pitch mounts pivot the display up and down. Swing mounts articulate up and down, and side-to-side.
- **Projector:** Designed for projectors, these mounts are generally universal-style mounts that accept a lot of different projector models. This base mount unit is then complemented with other accessories for attachment to a ceiling, wall, pole, or other mount. If you have a drop ceiling, for instance, you would attach the projector mount to a variable-length pole that allows you to make your mount installation flush with the ceiling tiles.
- **Automated:** Designed for applications where you want to hide your gear when not in use, automated mounts lift or drop your HDTV into the ceiling, floor, or cabinet. Click a button, and voilá, your TV is back. Automated swing mounts can, with the press of a button, swing your TV into a preprogrammed position.

You can install many mounts yourself. However, if you're thinking about an automated mounting solution, plan to hire someone. A whole different level of expertise is typically required to manage the environmental, electrical, and other design elements of a successful implementation. If you mess up this one, you could end up dropping an expensive TV (ouch) or burning the house down (double ouch!).

Buying Your Mount

You can get basic mounts at your local Costco, Sam's Club, or other HDTV dealer. You can also buy them online from places like Mounts Direct (`www.mountsdirect.com`).

We shudder when we see people buy a great TV and a cheap mount. That's a recipe for a very lonely and sad weekend when it falls off the wall. A good mount can last you through many TV displays, so make the right investment and get a good one. Don't let cost be the only determinant in your buying decision. Call some stores and see which ones the employees like.

Consider these key things when buying a mount:

- **Great design:** We like mounts that are rather minimalist in their design. A well-designed mount doesn't weigh a lot or take up a lot of wall space. The area that actually attaches to the wall should be small, leaving you with more space behind your TV for cable and electrical outlets.

  Your electrical and networking wall outlets were probably installed before you got your wall mount. A sign of a poorly designed wall mount is if it obscures one of these outlets because the surface of the wall plate is too big. We really hate it when that happens because there's little you can do about it except return the mount.

- **Ease of installation:** Look for hands-free installation of the wall mount. One person can install a well-designed mount. For example, the Chief Manufacturing (`www.chief.com`) flat-panel mounts allow you to pound a nail in the wall and temporarily mount the whole unit on that one nail, leaving your hands free for leveling and final bolting. If your mount is heavy — they can weigh up to 50 pounds — you want to be able to pre-install the top two lag bolts and then slide the mount over those bolts, leaving the remaining bolts easier to install.

  Some mounts come with *integrated lateral shift capability.* This capability lets you install a mount off center so you aren't tied to the spacing of the studs in your walls. After all, it would be just dumb luck if the spacing of the studs happens to perfectly coincide with where you want to center your display. A high-quality mount allows you to displace your mount at least 4 inches to the left or right with 16-inch or 24-inch centered studs. You should be able to vertically position an installed display as well; quality mounts allow for at least 3 inches of vertical shift.

- **Ease of use:** You should be able to move a great mount with the tap of your finger, and it should stay in place without you having to tighten all sorts of knobs. You'll notice this capability if you swivel or tilt your display a lot — you don't want to be messing around with all sorts of settings.
- **Ability to move (tilt, swivel, extension):** Based on your mounting plan, check the planned field of movement with the technical specifications of the mount. Remember that different-sized screens can achieve different motion, depending on clearance with the wall or ceiling when installed.
- **Rated weight capacity:** Know the weight of your display *plus* any accessories — like speakers, decorative trim, artwork screens, and other accessories — that can add weight. This is especially true with the lower end of the market where the margin of error is not as generous.
- **Ability to integrate cable management:** No one likes to see a beautiful HDTV display with cables dangling all over the place. Your cables should become an integrated part of your mount. Understand how your wiring runs through the mount. This is especially true for ceiling mounts for projectors where the cabling might go through the center of the mount piping. If you're mounting a projector into a cement ceiling, you can have problems getting the cable out of the pipe with a flush mount. Also check for snap-on covers if you're running cables outside of your mount.
- **Ability to detach/reattach at will:** Good mounts allow you to easily detach the HDTV without having to uninstall the mount. For instance, if you ever have to replace a light bulb in your projector, you want a mount with a quick detach capability so you don't have to reset and reconfigure your projector when you put it back.
- **Great aesthetics:** Mounts come in all shapes, sizes, and colors. Get exactly what you want. Look for a low-profile mount. Check for aesthetic details, such as covers for lag bolts on the wall. The better-quality mounts have these important extras.

Installing Your Mount

Because so many mounts are available, it's hard to tell you exactly step by step what to do. However, we can certainly guide you in the right direction with some time-honored lessons learned.

- **It's all about the anchors.** Great mount, wrong anchors, bad results. We cringe every time we hear those stories. Take the time to figure out the right anchors for your job, or you'll be very sorry. There are specific anchors for specific types of materials. Don't be the local tech blog's funny story of the day about the guy who tried to mount a 50-inch plasma by using drywall screws.

Most mounting instructions start with the assumption of wood stud installations, but there are specialty kits (with some mounts) for concrete anchors, drywall anchors, or metal studs. A good rule to follow is that your anchors should hold five times the weight of what you're putting on the mount. Most manufacturers test their products with 5⁄16-inch lag bolts (some use 3⁄8-inch) although a 2-1⁄2-inch-long lag bolt is generally acknowledged to be acceptable for large mounts. The main goal is to hit the center of the stud with the bolt. Use a stud finder, or — if you must — drill pilot holes to find the width of the stud so you can determine its exact center before you actually put in your bolts.

- **Double-check all clearance requirements.** Most mounts require that you lower the display onto the mount. This means you need to have a little more clearance above the mount to make sure you can slide the display into place. Generally, this can range from 1⁄2 inch to 4 inches, depending on the manufacturer, so if you have a really tight space, think about this in advance. Never plan for an exact fit, or you'll be sorry.
- **Think of the complete range of motion of the fully mounted display.** If you have designed your mount to tilt, swivel, or extend your display, then you need to consider the full range of motion for the mount plus the display as an installed unit, plus any other accessories like speakers. As a completed unit, it might not have the full range of motion promised on the mount box, because of sheer clearance issues specific to your installation. We've seen instances where someone bought a mount with the expectation of a 45-degree angle, only to find that, because of the size of the display and the limited extension from the wall, the full 45 degrees could not be achieved.
- **Consider protecting your lifted displays.** When you lift (or drop) your display by using an automated mounting system, your display is only as protected as the space where you store it. If your storage space is an attic that isn't insulated, your display is exposed to the same elements your attic is, including extremes of temperature. Projector lifts, on the other hand, tend to be enclosed in a box that has plenum rating and can go in a false ceiling.
- **Check your wall structure to make sure it can hold the weight.** Your walls might not be strong enough for the HDTV you want to hang on them. This is especially true now that the larger wall-mountable LCD and plasma units are getting cheaper. The bigger the display, the more likely you might need to create a special mounting structure within the wall to support the mount. You probably need to take some wallboard or plaster off.
- **Check and double-check your measurements (and then ask your spouse to check it again for you).** It pays to take time and practice the field of movement with the display, as if it was mounted, to ensure that everything fits just right. Having good lateral and vertical shift helps, but that can cover up only the most minor of mistakes.

Don't feel like you failed if you call in a professional installer to help you mount your HDTV. This is not like asking for directions if you get lost . . . well, actually it is, because you should do that, too! Just know this is one area where, if you need to, you should pay the money to get it done right. A good mount and installation should last you many TV sets. The mount shouldn't keep falling off the wall.

Disguising Your Display

Sometimes you just don't want your TV to look like a TV. Whether you have that special living room where it just doesn't look right or a spouse that is just pathologically opposed to new technology, you can get some accessories for your mounted display that can help.

For instance, you can get old-style picture frames that make a display look more like a painting on the wall than a TV. You complement this frame with your favorite digital art. Connect your PC to your display via GalleryPlayer Media Networks (`www.galleryplayer.com`), and you can virtually tour your favorite museum paintings — on your framed HDTV screen (see Figure 5-2).

Figure 5-2: Art for your HDTV, from Gallery-Player.

GalleryPlayer is a digital home "gallery service" that delivers museum-quality art and photography to any high-definition display, transforming it into a digital canvas. The application is free, and you can display hundreds of free images, as well as browse from individual pictures and collections that turn your screen into an art gallery. Images cost less than a $1 each, so check it out.

Don't display just one painting if you have a display that burns in images, such as plasma displays. Cycle through paintings to even out the color usage across the screen. Burn-in *can* cause permanent degradation of your screen (though most modern plasmas handle fixed images pretty well), so you should make sure you change images often!

Alternatively, you can purchase a masking system that hides your screen altogether. For instance, BEI Audio/Visual Products (`www.beionline.com`) offers the BEI Motorized Artwork System, designed for raising and lowering a framed canvas artwork to conceal or reveal a plasma or LCD display. Just tap a button, and the canvas slides away, and your favorite episode of *SpongeBob SquarePants* comes on. (Pat's a fan of ole SpongeBob.)

The BEI system offers more than 300 reproductions of museum master works of art — Monet, Manet, Millet . . . you have a broad selection of the most popular works of art. You pick from a range of more than a dozen classic wooden frames to match your décor. A BEI system costs around $2,500; the option of bulletproof glass is extra. (Yikes!)

Part II
Getting HDTV Programming

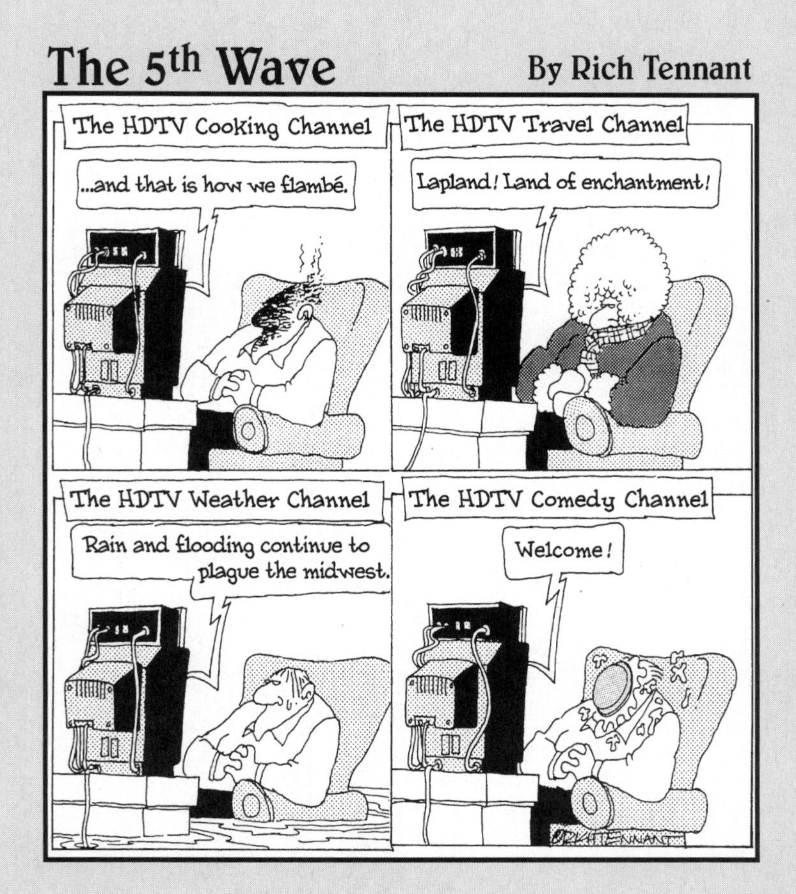

In this part . . .

You probably want to actually look at an HDTV program, not an inert HDTV set. A couple of years ago, there were none to be found, but today you can find an ever-increasing amount of HDTV programming available.

In this part, we expose you to various ways you can get high-definition content onto your HDTV. We start with a general look at the signals available from the over-the-air broadcasters, cable companies, and satellite companies. We reveal why not all signals look the same from these sources.

Then we drill down on each of the three major originators of HDTV signals, from the over-the-air broadcasters to cable companies to satellite firms. We tell you about the HDTV antennae and how to optimize your OTA connections. Then we look at how the cable and satellite firms are delivering high-definition cable programming. We tell you about the role of your analog and digital set-top boxes as well as how to upgrade to HD satellite service if you're a customer, and even how to combine satellite service with Internet access.

We wrap up our look at HDTV programming with a deep dive into the burgeoning world of Internet-based content. Did you know that you have the option to watch HDTV movies off the Internet? Read on to find out how.

Chapter 6

Who's Showing HDTV?

In This Chapter

- Deciphering HDTV providers
- Checking out free HDTV!
- Getting into pay TV options
- Specializing in HDTV
- Measuring your bits and bytes

In the vein of "all dressed up and nowhere to go," HDTVs would be no good without content to show on them — native, high-definition content, that is. Hollywood has done its usual: "If there's a place to put the content, we'll come out with it." And the manufacturers have said, "If you make the content, we'll build the systems." Lucky for you, DVDs and higher-definition camcorders have created a lot of reasons to want higher-resolution TVs, and that has prompted the broadcasters to follow suit with their own HDTV content.

Broadcast TV stations, for example, are heeding the FCC (Federal Communications Commission) and offering plenty of HDTV broadcasts. Cable companies are offering at least some HDTV programming (and usually not just "some" programming but "a lot") in every single one of their major markets. And satellite providers are offering HDTV to pretty much the entire United States — so almost no one is outside the reach of HDTV services these days.

In this chapter, we take a high-level look at who's offering HDTV, what it takes to get the service, and what kind of HDTV content you can get from the different providers. That last item can be important to know if, for example, you bought your HDTV specifically to watch *The Final Four* (we predict a big Duke run next year!) or for seeing Tony and Carmela in the utmost detail in the last season of *The Sopranos.*

In the three chapters that follow in this part of the book, we get into a lot more detail about broadcast, cable, and satellite HDTV — the programming, its availability, and the equipment requirements. Consider the discussion in this chapter a quick, up-front primer. Move forward in the book when you're looking for the details.

Looking at Who Has HDTV

We're in the midst of HDTV-mania. If you happened to have a dead cat handy (and no, we're not cat haters; we love cats and don't want them dead), you wouldn't be able to swing it very far before you hit an ad for HDTV sets, HDTV programming, HDTV services from a cable company, or something HDTV.

This situation makes us happy. We've been going to the CES (the Consumer Electronics Show, the biggest HDTV-related trade show there is) every year for a long time. And every year we hear, "This year, *this year,* is the Year of HDTV!" Well, we've finally reached that point — in what must be the fifth or sixth official Year of HDTV (YOH for short). Thank goodness.

We know that it's now YOH because of the sheer volume of HDTV programming that's become available from different sources and because of the profusion of ways the average person can now access that programming.

Broadcasters

The TV providers who have begun to really see the light are the broadcast networks (such as ABC and CBS) and local affiliate stations that broadcast the network content over the airwaves. This hasn't happened purely because the broadcast folks are being good TV citizens. It's because the FCC has mandated a transition from analog to digital TV. Eventually, all of these broadcasters will need to turn off their analog signals and send out DTV broadcasts. (Note that we say DTV, not HDTV — lower resolution 480p or even 480i signals can be broadcast over DTV.)

The date for this final transition has been changed a bunch of times by the FCC and Congress — essentially over concerns that folks won't be able to afford new TVs or external tuners to pick up TV broadcasts when the analog ones are finally turned off. The latest, and hopefully final, date for this is February 18, 2009. On that day, analog TV will be dead, and all broadcasts will be digital only. Why such a weird date? Well it's after the Super Bowl, so people won't have to worry about forgetting to get a digital TV or tuner and therefore missing the biggest broadcast event of the year.

It's not all pressure from the FCC, however, that is driving the switch from analog to digital. Broadcast networks are increasingly competing with what we call *cable networks.* Though it's a bit of a misnomer, we use this term to describe networks you can get only via cable or satellite (such as HBO or TNT). Broadcasting in HDTV — especially for big-ticket items like prime-time shows and major-league and college sporting events — gives the broadcast networks a leg up on cable networks.

All five major broadcast networks (ABC, CBS, NBC, Fox, and CW) are broadcasting at least some HDTV content through their local affiliates, though much of the programming remains in standard definition. If you tune in to the digital broadcast of your local channels during primetime, chances are better that you get an HDTV show!

The best thing about broadcast network HDTV (often called *OTA,* or over-the-air) is that it's free. Free as in *free beer.* No cost to you (except maybe having to watch some bad ads). All you need is an HDTV with an antenna and a built-in or external tuner. In Chapter 7, we give you some more detail on these accessories.

Cable and satellite networks

The cable networks (the networks you can't get with a rabbit-ear antenna) have not been sitting around idly while broadcast networks have been sending out free HDTV to their viewers. In fact, many cable networks have developed and launched their own high-definition channels, ranging from movie channels, like HBO or Showtime High Def, to sports, like ESPN-HD and ESPN2-HD.

The biggest problem that cable networks have had is finding cable or satellite networks to carry these channels to customers. To understand why, skip ahead to the section titled "All HDTV Signals Are Not Equal." Go ahead — we'll wait right here for a moment.

Basically (in case you didn't skip ahead), cable and satellite systems have a limited amount of *bandwidth* (or slots for TV channels) within their broadcast systems. HDTV uses five to eight times as much bandwidth per channel as does analog TV — or to reverse that, you can fit five to eight analog TV channels in the slot occupied by one HDTV channel. So cable and satellite companies have taken a while to begin to show interest in carrying HDTV channels. They thought they could make more money off eight analog channels of fly-fishing and underwater basket-weaving than they could from one HDTV broadcast of *The Final Four.*

The good news is that cable and satellite providers have begun to carry at least a handful of HDTV stations — often more. To get into these HDTV cable signals, you need the following:

- An HDTV service contract with your cable or satellite company. Unlike broadcast, this service isn't free — you have to pay the piper.
- An HDTV satellite receiver or cable set-top box.

A digital-cable-ready TV and a CableCARD, which are becoming relatively widely available, let you skip the set-top box. See Chapter 8 for more details.

Specialized HDTV stations

Existing broadcast and cable networks aren't the only ones to realize the potential of HDTV. A small flurry of new networks specifically delivers HDTV channels to cable and satellite providers.

A good example of this is HDNet (`www.hd.net`), the brainchild of Dallas Mavericks owner Mark Cuban. HDNet was launched with the sole purpose of providing a range of original programming, such as news, sports, and series, along with licensed programming from other studios, such as Andy Richter's show, all in full 1080i HDTV. Danny is particularly interested in HDNet's show "Bikini Destinations." (Pat made that up!) You can get HDNet's two channels on both of the major satellite networks, as well as a growing number of cable provider's networks.

HDNet was the first — but not the last — high-def-only network. For example, a cable-only network owned by a consortium of cable companies called INHD (`www.inhd.com`) provides two channels with a variety of original and licensed programming of sports, movies, and other content. Hey, that sounds familiar, huh?

Making Your Choice

We have some good news and some bad news. The good news is that you probably have a choice of where you get your HDTV content. The bad news is, well, you have to choose. If you're like us, making your choice means hours of research and poring over Web sites, trying to figure out what works best to fit your HDTV needs.

Well, we don't have a magic bullet for you. Because HDTV availability is highly dependent upon *exactly* where you live (right down to the street address — it can differ even within neighborhoods), we can't give anything but the most general advice.

There's nothing wrong with mixing and matching amongst these different sources. For example, if you want local HDTV content along with your satellite-TV source, you need to hook up an antenna to your dish and pick up the OTA broadcasts. Luckily, most HDTV satellite receivers have a built-in OTA HDTV tuner, so you don't need extra equipment (beyond the antenna).

So, given that wishy-washy disclaimer (sorry, but it's true!), here's our official advice:

- **Figure out what's even available.** In each of the following three chapters, we give you some pointers to Web sites and other resources that help you find out what you can get in your house.
- **Look at your budget.** Keep in mind the fact that "free" OTA HDTV might not be free if you have an HDTV-ready system and need to spend hundreds of dollars on an external HDTV tuner. Cable, on the other hand, might include a monthly fee but doesn't require any up-front expenses for tuners or set-top boxes. Many cable companies give you local HDTV channels free for the price of the set-top box rental ($10/month or less, typically). Satellite might have lower monthly fees than cable, but it also requires an up-front purchase of the receiver.
- **Examine closely the channel lineups available to you.** Remember that quantity and quality are two different things. For example, a satellite company like DIRECTV might offer more HDTV channels than your local cable company, but you might not be interested in watching all of them, and you might not be able to get your local network affiliates in HDTV on the satellite system. For example, do you really want to spend your HDTV viewing time watching some nature special on the mating habits of frogs instead of watching *The Unit* in HDTV on your local CBS affiliate? Look for the channels that you love.
- **Consider the performance.** In the following section, "All HDTV Signals Are Not Equal," we discuss some of the ways that various TV providers throttle back their HDTV signals to save bandwidth on their networks.

Going online for HDTV

Traditional broadcasters aren't the only folks who can send HDTV broadcasts and programs into your home. A growing number of providers offer HDTV (and standard definition, too) over the Internet. In Chapter 10, we dig into the details, but here's the high-level view: You can connect a PC (a home-theater PC, discussed in Chapter 16) or a specialized Internet set-top box (we discuss these in Chapter 10) to access HDTV programming that's available for free or for a fee online. For the vast majority of folks, this kind of TV programming doesn't replace one of the other systems (OTA, cable, or satellite), but rather *augments* them, just as high-definition DVDs or tapes augment what you watch on your HDTV. Over time, we fully expect that you'll get everything you need over the Internet, but right now there just isn't enough content nor enough *bandwidth* (or Internet connection speed) in the average home to make the Internet a full replacement for the more traditional services.

A variation of this is *IPTV* (or Internet Protocol TV), which uses the same underlying network mechanisms as the Internet, but is more like a cable TV offering that just uses different wires. We talk about IPTV in Chapter 10 as well.

In the end, you might find something that we haven't mentioned or something less than coldly rational and logical (Mr. Spock to the bridge!) that makes you decide on an HDTV provider. For example, for Pat (actually, for his wife, who's a huge baseball fan), it was a no-brainer: He wanted satellite, but then his local cable company started broadcasting all of those San Diego Padres games in HDTV. Case closed — cable it is for Pat.

All HDTV Signals Are Not Equal

Not all HDTV signals are equal across all providers. When the signal is sent from its source to you, it is usually compressed to cut costs and bandwidth requirements. How much the signal is compressed — and what digital encoding scheme is used — determines a lot about what you see on your HDTV.

MPEG-2 is the standard used by digital-TV broadcasters today to compress, encode, and then ultimately decode the TV programs. Compression is necessary because a limited amount of space is available on the cable, satellite, and broadcast-TV networks for transmitting these signals. Table 6-1 shows you how the bandwidth is allocated.

Table 6-1 Who Has the Most Room for HDTV?

	Terrestrial Broadcast	*Satellite*	*Satellite*	*Cable*	*Cable*
Bandwidth	6 MHz	24 MHz	36 MHz	6 MHz	6 MHz
Modulation	8 VSB	QPSK	QPSK	64 QAM	256 QAM
Bit Rate	19.39 Mbps	27 Mbps	40.44 Mbps	27.7 Mbps	38.8 Mbps

Source: Ultimate AV Magazine (June 2004)

Suppose an over-the-air broadcast TV signal starts out its life at the central network hub as a 1920-x-1080i signal. This signal can first be encoded for broadcast at a rate of around 995 Mbps. By the time it is sent to you over the air, it's compressed to a mere 18 Mbps signal. That's a lot of compression, but the picture still looks great.

The station could choose to compress that signal a little more, say down to 13 or 14 Mbps to make room for other channels in the same signal. The difference between an 18 Mbps and a 14 Mbps signal is indeed noticeable, especially when the programming contains a lot of motion. And don't forget, once the signal enters your HDTV set, it probably encounters *another* round of downward resolution as it tries to put the original 1080i image on, say, a 720p TV set.

Satellite providers also have similar compression challenges. *Transponder space* (the satellite-located transmission systems) is so expensive that compression is required for everything. Popular channels are typically encoded at 15 Mbps or more before they go to the satellite operators. Depending on the available satellite bandwidth (could be 24 MHz or 36 MHz, which yields 27 Mbps and 40.44 Mbps of bandwidth respectively), satellite operators can compress these 15 Mbps signals down to 13.5 Mbps, so they can cram two or three HDTV streams on these satellite signals, respectively.

The satellite guys are launching a bunch of new satellites and converting their broadcasts to *MPEG-4*. This is a newer, better, and more efficient encoding/compression system than the MPEG-2 now in use, which will allow them to have *the same level of picture quality* using less bandwidth. In Chapter 9, we talk about this transition, when it will happen, and how it affects you. (You'll most likely need a new satellite receiver.)

Cable operators have about the same options with the two major digital cable modulation schemes in use today: 64 QAM, which offers a maximum bandwidth of 27.7 Mbps, and 256 QAM, which has 38.8 Mbps available. They can compress the signals further and cram more channels into each signal, or they can offer higher-quality signals as a way of competing better in the market. After all, the 38.8 Mbps rate of the cable companies is twice the bandwidth per signal compared to the 19.4 Mbps data rate of the broadcast TV folks — so they can send two HDTV streams for each single HDTV stream that the over-the-air stations can send.

Who's required to do what?

Your over-the-air digital TV broadcaster has only 6 MHz (19.39 Mbps) of bandwidth over which it can send its TV signals. It can fill this bandwidth in any way it chooses, combining such broadcasts as HDTV, SDTV, weather images, and even FM broadcasts onto one bandwidth stream. The FCC, through its DTV rules, has said broadcasters must transmit one standard-definition digital-TV signal — 480i — and has not said anything about requiring HDTV.

So TV stations have to make some economic decisions about how best to use that signal. In one chunk of 6 MHz bandwidth, a broadcaster could send one full-quality, 720p 60-frame-per-second (fps) HDTV program; or two shows, one (say) for a 720p or 1080i HDTV show and one 480i SDTV show, both at 30 fps; or four 480i SDTV 30 fps shows. How a station mixes and matches its signals depends as much on its technology as its positioning in the market.

Cable companies have to deal with the sheer number of places where the signals are converted from digital to analog and back. Signals start out as digital, but the distributor might convert them to analog. The recipient cable operator then converts the signal to a *digital composite* (that is, component) signal. Then, as it's fed into the TV display unit, the signal is converted to one of several types of video signal: analog component, S-video, or composite. TVs that take in digital signals via digital interfaces can reduce some of this conversion — as can improvements in signal distribution between the broadcasters and the operators.

So, as complicated as this is, what does it all mean for the average buyer? Here's the lowdown for you:

- **Not all HDTV signals are going to be the same.** It depends on how the broadcasters compress the signals for the stations that you want to watch. Channels with a lot of action (like sports) tend to get more bandwidth; channels with a lot of static pixels (like the Home Shopping Network or any cooking channel) can survive more compression.
- **If you plan on watching a few channels a lot, do some research.** Your ultimate goal is to find out how much bandwidth these channels are using. You can't easily measure this yourself (and it's about impossible without installing the service, if you're using cable or satellite), but try to find a local bulletin board or newsgroup online to see what the local HDTV nerds have to say. For example, Pat spends a lot of time on `http://hdtv.forsandiego.com`, reading about experiences with local HDTV options. You can find enthusiastic, smart folks who have the tools and the know-how to examine HDTV bit streams, measure them, and come out with concrete evaluations. Also, watch your neighbors' HDTVs now and then and compare what you can — if it *looks* different, it probably is.

Chapter 7

Something's in the Air

In This Chapter

- Getting HDTV broadcasts for free!
- Finding out who's broadcasting HDTV
- Getting equipped for OTA
- Choosing an antenna
- Picking up signals

In most ways, HDTV is an amazing leap into the future. It's digital! It's high-definition! It's shiny and new! It's expensive! (Well, it doesn't have to be *that* expensive, but HDTVs do cost more than old-fashioned analog TVs.)

But HDTV does have one aspect that brings you back to the olden days of black-and-white TVs — the antenna has reappeared! And with the antenna comes all the fun of Dad standing on one foot and twisting the rabbit ears *just so* to bring in that hard-to-tune channel.

Okay, so it's not that bad (most of the time), but it can be challenging to tune in to over-the-air (OTA) broadcast TV. We're not going to kid you here — cable-TV or satellite setups are much easier to use than an antenna.

But OTA HDTV is free. FREE! And nearly every household in America is in range of at least one of these free HDTV stations (99 percent of homes, according to some sources).

In this chapter, we discuss how you can figure out which OTA HDTV signals are available in your location (there are some great online tools!). Then we help you pick out the equipment you need (an antenna and a tuner), and finally we tell you how to get it all set up.

Finding Local HD Broadcasts

Before you do anything else, do your homework. In the case of OTA HDTV, you need to simply spend a little bit of time online looking for broadcast HDTV channels. You don't have to do this research online, but if you have Internet access, that's the best way to do so.

If you don't have Internet access, talk to the retailers from whom you're buying your HDTV — they probably have firsthand knowledge (or at least anecdotal information) about local HDTV signal availability. They might also have a kiosk that provides access to the same online information we talk about in this section.

If you do have Internet access, you can find an extremely easy-to-use HDTV *signal finder* online. A company called Decisionmark, which sells software to television broadcasters, has created an online system called TitanTV that lets you enter your address information, press a button, and come out the other side with a nice listing of all your HDTV-channel choices (and all of your non-HDTV channels, as well).

Just go to `www.titantv.com` and follow the on-screen instructions. You can also access TitanTV at many HDTV retailers and come home with a printout of your available stations.

Not only does TitanTV find OTA stations for you, it also comes up with the cable and satellite stations available to you. You can do a quick comparison and see what best meets your needs.

The Consumer Electronics Association (CEA) is the huge (and hugely influential) trade organization that includes just about every HDTV manufacturer in the world. So it has a vested interest in getting people to buy HDTVs. This interest is sometimes manifested as lobbying efforts with the FCC, or efforts to develop marketing and industry. It also pops its head up in the form of direct-to-the-consumer education efforts.

So the CEA teamed up with the folks at Decisionmark to provide you with another way to view a listing of local HDTV broadcasters with the AntennaWeb.org system, which we discuss later in this chapter, in the section titled "Choosing the right antenna." AntennaWeb.org is basically the same underlying database used by TitanTV, optimized to help you choose an antenna — TitanTV is designed to help you see what shows are being broadcast.

Tuning In

HDTV breaks the old TV paradigm in many ways. One big difference is that not all HDTVs have a built-in TV tuner. Back in the olden days of NTSC and analog TV, TVs without tuners were quite rare. A few high-end TVs were monitors and required an external tuner, but they were definitely in the minority.

With HDTV and OTA HDTV broadcasts, the situation has been reversed — at least for the time being (more on that soon). Many sets sold as HDTVs today are *HDTV-ready* but don't have any electronics inside them that can pick up an OTA HDTV broadcast (or *any* ATSC broadcast, whether high definition or standard definition — and if you have no idea what that means, check out Chapter 1 for definitions of ATSC and NTSC).

The government has begun requiring TV manufacturers to include built-in tuners that can pick up OTA HDTV. The rollout started with bigger (36+ inches) TVs in 2005 and will eventually filter down to smaller sizes. By March 2007, all new TVs should be capable of picking up ATSC signals — even if they are *not* HDTVs and can't display the higher HDTV resolutions.

Building on a built-in tuner

If you have an HDTV with a built-in ATSC tuner, you're just about all set. All you need to do is find the appropriate antenna (see the following section "Antennas A to Z"), make the connections, and go. It's really that simple — or at least it can be.

Just follow the instructions in your HDTV's manual for tuning in the HDTV stations — we can't help you there because each HDTV on-screen setup process is different.

Some HDTVs with built-in ATSC tuners also have special tuners that can decode *QAM*-encoded HDTV signals. QAM is the system used by most cable TV networks. This means you might be able to pick up your local broadcast stations by just plugging in your cable TV connection. Note that this is different from the DCR (digital-cable-ready) systems we discuss in Chapter 8. And no, this has nothing to do with using an OTA antenna, but we thought it was a nice morsel of information to know in case you're ever asked this question at a neighborhood BBQ.

Adding on a tuner

If you own an HDTV, chances are good that you won't have the built-in ATSC tuner you need to pick up HDTV broadcasts. Your HDTV probably has an NTSC tuner, which can pick up analog broadcasts, but it probably is only HDTV-ready, so you need to pick up an HDTV tuner box that you can connect between your antenna feed and your HDTV.

The biggest problem with HDTV tuners (whether they're in your TV, or external) has traditionally been an economic one. They are (or at least were) darned expensive. Just a few years ago, HDTV tuners commonly cost $1,000 or more, which effectively explains why so many HDTVs were sold as HDTV-*ready:* Not everyone with an HDTV uses the OTA tuner, so why drive the cost of an already-expensive HDTV through the roof?

The good news is that prices have come down. Way down. As we write (summer of 2006), you can buy a name-brand (Samsung, in this case) HDTV tuner for about $250. That's less than six months' worth of double espressos at Pat's favorite café! And the prices are going nowhere but down.

When you're choosing an HDTV tuner, here are just a few things to consider:

- **Digital outputs:** If at all possible, use a digital cable connection between your tuner and your HDTV: DVI-D and HDMI (see Chapter 3) are by far the most common. (1394/FireWire used to be common, but it's rare these days.) Make sure that the outputs of your tuner match the inputs of your HDTV.

  The newer HDMI system is *backward-compatible* with DVI-D with the use of a simple adapter. So you can mix and match DVI-D and HDMI freely.

  If your HDTV tuner's DVI-D or HDMI output uses the HDCP copy-protection scheme, make sure the DVI-D or HDMI input on your HDTV does, too. Otherwise, the system might convert the signal to a lower resolution, giving you a non-HDTV picture.

- **Analog outputs:** Although you want to use your digital outputs, if possible, it's handy to have a full set of analog outputs on the HDTV tuner, for making connections to other devices (like a DVR).
- **Output resolution:** You can adjust most HDTV tuners to match the best resolution for your HDTV. Some HDTVs require a specific signal resolution (such as 1080i); if yours does, make sure your tuner can give you output at that resolution. It's rare that an HDTV tuner sold today won't offer you a choice of either 720p or 1080i, so pretty much every tuner works with every current HDTV's inputs. Only if you're picking up an older unit on eBay or Craigslist.com do you need to worry about this.

- **Satellite capability:** Some OTA HDTV tuners also include satellite TV receivers (for services like DIRECTV or DISH Network, see Chapter 9). Well, the satellite companies would flip it around and say their receivers include OTA tuners. Either way, this can be handy if you're using a satellite service for premium HDTV channels (like ESPN-HD and HBO) and using an antenna to pick up local HDTV channels.

As we mention in the previous section, the FCC now requires TV manufacturers to begin including built-in ATSC tuners in their TVs, as part of the overall industry transition to digital TV. But many folks who are buying flat-panel (plasma or LCD) displays or front-projection systems are buying displays that are monitors, not TVs — in other words, they don't have *any* built-in TV tuners. In these cases, the legal requirements don't matter, and you need an OTA tuner, a cable box (see Chapter 8), or a satellite receiver (see Chapter 9) to get broadcast HDTV signals into your TV.

Antennas A to Z

The other half of the OTA HDTV equation is the antenna. Yep, the rabbit ears are back. Well, you probably don't want to bring back some of those unwieldy old antennas of years past, because the TV antenna has gotten in shape for the new millennium.

Choosing the right antenna

You probably *could* use those old rabbit ears — there's nothing that incredibly unique about picking up HDTV broadcasts — but tuning them in (which is what the HDTV tuner does) is a complex process. Luckily, a lot of great new antennas on the market work particularly well with HDTV broadcasts.

Here are a few antenna features to consider:

- **Location:** Inside the house or outside? Well . . .
  - *Outdoor antennas* work better, but they also require work, space, and access outside your home.
  - *Indoor antennas* might work well enough for you if you're close enough to the transmission towers.
- **Direction:** Where to point the thing? Look here . . .
  - *Directional antennas* can be aimed toward a specific point (or within a certain angular vector).

Directional antennas usually are better at picking up distant signals and can reduce *multipath* distortion (what happens when the signal bounces off buildings and other structures, and you receive the signal multiple times).

- *Multidirectional antennas* pick up signals coming from — you guessed it — multiple directions.

  You can mount multidirectional antennas just about anywhere, and pick up signals from widely dispersed transmission towers.

- **Amplification:** Some antennas have a built-in *amplifier* that increases the strength of weak broadcast signals before they get to your tuner.

  Amplification is generally a good thing, unless you are close to the broadcast antenna — you don't want to overamplify a signal.

It's really difficult for even a technically savvy person to determine which antenna is right for a particular home — you have to figure out a swarm of variables. That's why we're so happy that the CEA and Decisionmark (the folks we mention in the first part of this chapter) have come up with the AntennaWeb.org system. (That's also the URL: `http://antennaweb.org`.)

AntennaWeb.org is a Web-based database that "knows" your HDTV situation from your address, the location of HDTV broadcast antennas near you, and some sophisticated software that models HDTV-signal propagation in your area (by looking at various geographical and demographic data, such as the terrain and the presence of high-rise buildings). Just go to the Web site, enter your address, answer a few questions about your surroundings, and click the Choose an Antenna button.

Up comes a results page, as shown in Figure 7-1. The AntennaWeb.org system uses a graphical representation (a pie chart/color-code system) called the *antenna mark* to identify seven different antenna types with unique different colors and coverage zones. The pie chart is sort of like a Trivial Pursuit game piece — the more slices of the pie you have, the better. The coverage zones represent the footprint (or coverage areas) of the antenna based on its designs and use. All the HDTV broadcasts in your area that you can at least theoretically pick up are shown (as well as the analog stations), and each is color-coded.

To find the right antenna for you, simply note the color code for the stations you want to pick up and go shopping for an antenna that matches that color. You see a graphic on the box of the antenna corresponding to the one on the Web site. Click the color on the list on the AntennaWeb.org page to show the pie chart we discuss.

Please note how AntennaWeb.org fills in the antenna-mark pie chart — clockwise, starting with the yellow at 12 o'clock and moving around clockwise to the pink at 11 o'clock. Most antennas work for all the lesser — that is further back counterclockwise — color codes. It's common to find high-power directional antennas that also work fine as short-to-medium-range multidirectional antennas. So if your listed signal sources require antennae coded yellow and green (the second color after yellow in the scheme), you can buy a green-code antenna — it can also pick up the yellow-coded broadcast signals.

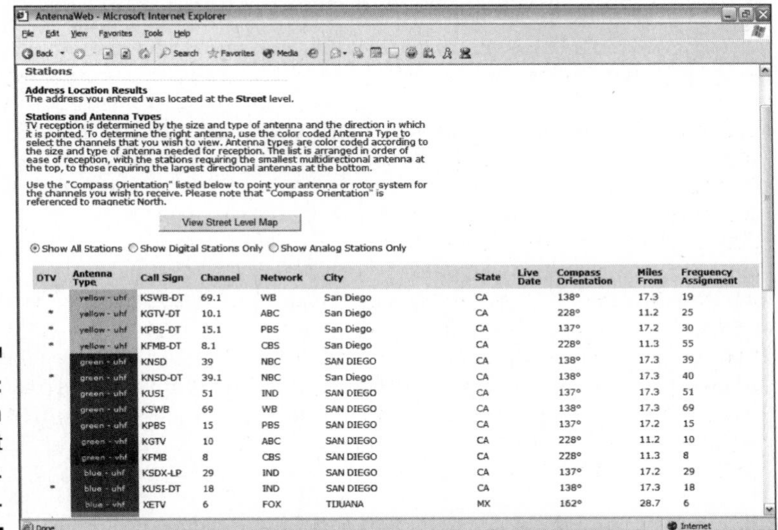

DTV	Antenna Type	Call Sign	Channel	Network	City	State	Live Date	Compass Orientation	Miles From	Frequency Assignment
*	yellow - uhf	KSWB-DT	69.1	WB	San Diego	CA		138°	17.3	19
*	yellow - uhf	KGTV-DT	10.1	ABC	San Diego	CA		228°	11.2	25
*	yellow - uhf	KPBS-DT	15.1	PBS	San Diego	CA		137°	17.2	30
*	yellow - uhf	KFMB-DT	8.1	CBS	San Diego	CA		228°	11.3	55
	green - uhf	KNSD	39	NBC	SAN DIEGO	CA		138°	17.3	39
*	green - uhf	KNSD-DT	39.1	NBC	San Diego	CA		138°	17.3	40
	green - uhf	KUSI	51	IND	SAN DIEGO	CA		137°	17.3	51
	green - uhf	KSWB	69	WB	SAN DIEGO	CA		138°	17.3	69
	green - uhf	KPBS	15	PBS	SAN DIEGO	CA		137°	17.2	15
	green - vhf	KGTV	10	ABC	SAN DIEGO	CA		228°	11.2	10
	green - vhf	KFMB	8	CBS	SAN DIEGO	CA		228°	11.3	8
	blue - uhf	KSDX-LP	29	IND	SAN DIEGO	CA		137°	17.2	29
*	blue - uhf	KUSI-DT	18	IND	SAN DIEGO	CA		138°	17.3	18
	blue - vhf	XETV	6	FOX	TIJUANA	MX		162°	28.7	6

Figure 7-1: Choosing an antenna at AntennaWeb.org.

Installing your HDTV antenna

When you've picked out your HDTV antenna, there's the simple matter of installing it. If you've selected an outdoor antenna (our recommendation for most users), you need to mount it.

Follow the directions included with the antenna — particularly any safety directions regarding such installation procedures as grounding and wiring. We want you to be safe and *not* burn your house to the ground for the sake of HDTV. (Not even we are that fanatical about HDTV.)

If you've chosen a directional antenna, part of the mounting process is aiming the antenna. The AntennaWeb.org results page actually includes (as you can see earlier in Figure 7-1) a compass bearing from your house to the antenna towers. So you can pull out your handy-dandy Boy or Girl Scout compass and do some aiming.

You can also use a *signal-strength meter* — many HDTVs or HDTV tuners have one built in — to determine how small adjustments in the antenna's position or direction affect your channels. Here's where it's handy to have a set of FRS (Family Radio Service) walkie-talkies, so you can communicate with the person in the living room who's viewing the meter while you're on the roof!

Tuning in to HDTV stations is a truly digital experience — either it works or it doesn't. For the most part, there's none of the fuzzy, half-visible pictures that you might remember from the old antenna days of analog TV. With the proper antenna, properly aimed, you might pick up all your local channels right away.

Or you might not. HDTV broadcasts are still a work-in-progress in many parts of the country. Sometimes it can be a hit-or-miss proposition — stations that should work don't, or they work only part of the time.

This situation is getting better every day, as broadcasters fine-tune their systems and as HDTV tuners become more sophisticated and adept at tuning in stations. As HDTV becomes more popular, many broadcasters are also turning up the power on their transmitters. (So far, they've been keeping the broadcast power down to save money!)

Chapter 8

The Cable Guy

In This Chapter

- Learning to love the cable guy
- Defining cable-service levels
- Figuring out set-top boxes
- Going analog
- Skipping the box altogether

Most Americans get their TV (NTSC or ATSC) from their local cable company — something on the order of two-thirds of all households subscribe to cable TV services. Cable companies have been extremely aggressive over the past ten years, rolling out new services like digital cable, cable modems, voice services (cable telephony), and video on demand (VoD).

In case you ever run into the term, cable companies are often referred to as *MSOs,* or multiple-system operators, because most cable companies own and operate cable systems in dozens or hundreds of different cities. Not a pretty term, but we didn't make it up, so we're off the hook!

More importantly, the cable companies have made a big push over the last couple of years to get HDTV programming onto their networks and into home theaters. Most cable operators now offer a range of high-definition TV services, ranging from pay movie channels like HBO, Showtime, and Cinemax to rebroadcasts of network HDTV stations (the local ABC, CBS, NBC, Fox, and other affiliates), and many have jumped in with some special HDTV-only programming (like HDNet, at `www.hd.net`). Many have even gotten into local HDTV programming. For example, in San Diego, where Pat lives, Cox and Time Warner (the two local cable companies) now broadcast over 140 of the San Diego Padres baseball games in beautiful HD! Go Padres!

In this chapter, we give you the basics on how to get HDTV from your cable company and what to expect when you do. We also tell you about some of the other services you can get from your cable company (things like digital cable, which are actually the majority of channels available on your HDTV, even though they aren't HDTV themselves). Finally, we talk about an exciting new trend: a new system of TVs and "smart cards" that let you forgo the cable box and plug directly into the cable coming out of your wall.

High-Definition Cable

The *crème de la crème* of cable services is HDTV over cable, and it's nearly everywhere. Cable companies now offer HDTV services in all of the 100 largest markets in the Unites States, according to the National Cable & Telecommunications Association, or NCTA (`www.ncta.com`), the largest industry group of cable operators. And even those who don't live in one of the largest markets can usually get HDTV from their cable system — 94 percent of the 210 cable markets defined by the NCTA have at least some HDTV programming.

Getting on the QAM bandwagon

The biggest difference between HDTV signals on a cable system and those broadcast over the air (see Chapter 7) revolves around the system used to modulate an HDTV signal for transmission as radio-frequency signals across the cable system.

Broadcast HDTV uses a system called 8-VSB (vestigial sideband), whereas cable usually uses a system called QAM (quadrature amplitude modulation). What's the difference? Well the technical specs are for the engineers. What's important to HDTV viewers is that you need a different kind of HDTV tuner to receive QAM signals than you do to receive 8-VSB (though many newer TVs are now including a tuner that can tune into both types of signals).

For most people, HDTV cable service comes through an HDTV *set-top box* (or cable box) that includes a QAM tuner. But a growing number of HDTVs have a built-in QAM tuner that allows users to simply plug the cable feed into the back of the TV and start receiving some of the HDTV programming offered by the cable company.

With *just* a QAM tuner in your TV, you can pick up only the free HDTV broadcasts (typically the local networks only). Later in the chapter, we talk about a device known as the CableCARD, which allows you to leverage a TV's built-in QAM tuner to receive the subscription-based HDTV services like HBO.

Encrypting and decrypting

The other big difference between broadcast and cable HDTV concerns *encryption* and premium cable channels. Broadcast HDTV channels — over-the-air broadcasts that you pick up with an antenna — are free for the picking (up). If you receive the channel, you can watch it.

Many cable TV HDTV channels, on the other hand, are considered premium or for-pay content — you have to pay extra every month to receive them. Because of this arrangement, most cable systems use an encryption system that scrambles the signal (using an encryption *algorithm* or formula). This means that even if you have a QAM tuner (whether built into your HDTV or attached to it), you can't watch these programs without some hardware to *decode the encryption* (in English translated from Geekese, that's *unscramble the picture*).

In some cable systems, the local broadcast HDTV channels are transmitted across the cable system unencrypted. If you have the appropriate tuner, you can view these channels without some sort of extra decryption hardware. We don't have an easy way for you to find this out ahead of time, but a simple call to your local cable company should net you the answer — just ask what HDTV channels are available without a cable box or CableCARD system. Keep in mind that the person you ask will probably try to up-sell you to a cable box or DVR, but you can always say no!

Although a very big exception to this rule is becoming pretty widely available on the market, the most common situation for most cable HDTV viewers is this: You need a set-top box from your cable company to view HDTV. This set-top box handles the QAM tuner duty, connects to your cable company to authorize you as a paying customer, and does the decryption/unscrambling of any premium channels. Figure 8-1 shows a typical HDTV cable set-top box.

Figure 8-1: Scientific Atlanta's 8300HD is the kind of set-top box your cable company might provide.

Pricing for HDTV set-top boxes varies widely by cable company. Some might charge you a monthly rental for the set-top box, but nothing for your HDTV channels. Others might not charge for the box rental but then charge you on a per-channel or package basis for HDTV channels. Still others might have a combination of these two approaches. In many cases, you're required to get the digital cable packages in order to get HDTV channels beyond ABC, CBS, and NBC. (Frankly, it's a mystery to us what your cable company offers, price-wise, so check out its Web site.)

Fee, fi, FiOS

Verizon, if you didn't know, is the largest phone company in the Unites States, serving most of the Northeast, and big chunks of Texas, California, and other states as well. If your local phone company is Verizon, you might have heard of something called *FiOS*. FiOS is Verizon's new fiber optic service bundle that uses (as you might guess) fiber optical cabling running right up to the side of your home. This fiber allows Verizon to provide phone service, faster (way faster) than DSL Internet service, and it also provides the infrastructure for a complete cable-like TV service.

And in fact, that is exactly what Verizon is offering customers in a growing part of its territory — over a million customers are now connected to this fiber optic network. We're not going to get into the details here, because from the customer point of view, FiOS is basically the same thing as a competing cable service — it's just delivered to your house differently. We mainly want you to be aware that this service is coming out and might be available where you live, so you can comparison shop between FiOS and cable (and satellite and other sources of HDTV programming).

Verizon is far from the only telephone company getting involved in the TV business. For example, AT&T (the second biggest telephone company in the United States) has been rolling out its TV service based on very fast DSL lines and using IP (Internet Protocol) to provide TV programming. We talk about these IPTV services in Chapter 10.

You might also find other folks offering fiber optic–based TV services as more and more homes are passed by these networks. In industry jargon, *passed* means that the network is in place in your neighborhood and they can turn your service on with a simple technician visit. For example, many municipalities and also developers are building their own fiber networks and offering cable-like TV services. So keep an eye out for the local advertising to see what's available in your area!

Your HDTV cable box typically also controls and enables all the cool (but non-HDTV) digital cable functions that you use on your HDTV — things like VoD (video on demand) and on-screen program guides.

You are not required to use a set-top box to receive subscription HDTV (and other digital cable) programming. Many folks do — either because their HDTV doesn't support the CableCARD system or because they want the set-top box for other features like program guides and VoD. See the section at the end of this chapter titled "Going Boxless" for more details on how to skip the cable box.

Digital Cable

The next notch down (picturewise) from cable HDTV is *digital cable*. This name has been the source of almost boundless confusion among many smart people we know — simply because the whole "digital" thing makes people think of DTV, and when you have *D*TV in your head, well then, *H*DTV follows.

We need to get this out of the way right away. Repeat after us: "Digital cable is NOT HDTV, digital cable is NOT HDTV. . . ." There, we got it out of our systems (and hopefully yours).

Defining digital

Now, we should hasten to add that in some places, HDTV channels might be part of (or an optional component of) a digital cable package. But that's just marketing.

Digital cable itself means two things:

- **Digital transmission and compression:** Standard-definition NTSC analog-TV channels are put through a digital wringer and come out the far end as digital channels (MPEG-2, typically, if you care to know the tech stuff) that take up less space on the cable system. In other words, you can cram more of them onto the cable system, so you can get the mythical 500 channels. These digital channels are then digitally transmitted by using a QAM system.
- **Two-way communications:** Digital cable utilizes two-way communications over the cable system, meaning the set-top box can send data upstream from your home to the cable office. Such two-way communication allows more sophisticated services like VoD to work, and that's why most cable companies began offering both digital cable and cable modems (which use the same upstream infrastructure) at around the same time.

Getting the benefits

So what does digital cable offer you (and your HDTV)? A lot actually (though, to repeat ourselves, HDTV isn't part of what's offered):

- **More channels:** Compressing each channel digitally allows more channels to be sent to your HDTV.
- **A high-quality NTSC picture:** Because channels are transmitted digitally, they are less prone to suffer from interference and "noise" than analog cable.
- **High-quality audio:** Stereo audio is provided with most programs; some channels might include a Dolby Digital surround-sound audio stream.
- **An on-screen program guide:** Most cable systems include an interactive program guide that lets you view about a week's worth of shows, view details about each program, and set up reminders and timers for recording.

- **Access to pay per view:** Unlike old-fashioned pay per view (PPV) programming, you don't need to call the cable company to watch a show (like a movie or sporting event). You can simply press a button on your remote to order.
- **Access to VoD:** The flagship service in any digital cable package is video on demand. This service uses the two-way network and on-screen guide, and allows you to select any movie or program from a library of programs retained on a server in the cable company's network — when you press Play on your remote, the program begins streaming down to your set-top box. You can pause, fast-forward, and rewind, just as you can with a DVD.

Getting the digital box

You need a digital cable set-top box in order to take advantage of many of these neat digital cable services. If you have an HDTV set-top box from your cable company, you probably already have a box that can take advantage of digital cable. (There are usually only a few dozen HDTV channels out of the hundreds of channels on any cable system; it makes no sense to build a box for HDTV alone.)

Like the HDTV-specific set-top boxes, digital cable set-top boxes are rented or leased to you (the customer) in a variety of different ways — such as a monthly box lease, a monthly service fee, or a combination of those. Look for set-top boxes that have extra features. For example, Pat is currently using a set-top box from Scientific Atlanta that supports HDTV and digital cable, and it has a built-in DVR (digital video recorder; see Chapter 12 for more info).

Analog Cable

The easiest cable signals to tune in are analog cable channels. Depending upon your cable system, all, none, or some of your cable channels are analog. (The latter case occurs when the first 50 or 100 channels on your system are analog, and all higher channels are digital.)

Analog cable is nothing more or less than NTSC, analog, standard-definition TV programming sent over cables instead of the airwaves. With the vast majority of HDTVs (that is, any HDTV that isn't just a monitor and has an NTSC tuner built-in), you can view analog-cable programming by simply plugging your cable feed (the RG6 cable and F connector) into the back of your TV.

If you're receiving digital cable or HDTV from your cable provider and are using a set-top box, you don't need to do anything special to tune in the analog channels. Your set-top box automatically tunes them in and sends them to your TV.

Some plasma TVs (Chapter 23) and most front-projection TV systems (Chapter 22) don't have a built-in NTSC tuner. If you're not using a set-top box, you can always use the NTSC tuner built into your VCR (VHS, S-VHS, and D-VHS VCRs all have NTSC tuners built-in) as your NTSC tuner.

Going Boxless

Like many other electronic devices and services, digital cable and HDTV cable are going plug-and-play. With the help of the government and industry organizations (such as the Consumer Electronics Association), cable TV providers have gotten together with technology vendors (TV manufacturers and the folks who build the infrastructure components of a cable network) to come up with a system that lets customers kick the cable box to the curb!

So what's going on? Two main things are happening:

- TVs are being built as *digital-cable-ready* (DCR) — just as regular analog TVs have been built as cable-ready for analog cable for years. These TVs have a QAM tuner built-in and also have a slot for a "smart card" called the *CableCARD.*
- Cable companies are getting ready to rent these CableCARDs to their customers. A customer with a CableCARD and a DCR HDTV can simply plug the CableCARD into the HDTV. The card basically gives your TV permission to unscramble encrypted premium channels. Typically, the monthly rental on a CableCARD is less than that for a set-top box — so you might pay $3 a month for the CableCARD (plus whatever your channel subscriptions cost) versus $5 or $10 a month for a set-top box rental.

This permission-giving process is called *conditional access* in the cable business. No, we won't test you on that. Just thought we'd share.

With a DCR HDTV and the CableCARD, you simply plug the cable into the back of your TV and you're ready to go.

The first-generation CableCARD devices (known as CableCARD 1.0) are one-way only — so they let you watch premium channels but can't communicate upstream to the cable company. This means that today's CableCARD deployments don't support digital cable services like program guides, pay per view, or VoD. Eventually, an updated, two-way CableCARD (CableCARD 2.0) will become available, allowing DCR HDTVs to access more of these advanced digital-cable features.

So is the set-top box dead?

As the CableCARD system gains widespread acceptance, the need for the set-top box will decrease. After all, who wants extra devices plugged into their HDTVs and sitting on the equipment rack if they don't need 'em?

Understandably, cable companies (and the set-top box manufacturers who support them) don't see it quite that way. Although CableCARDs are being offered (and will be required by the FCC to be an option for customers), the set-top box probably won't disappear completely.

Instead, the set-top box will become even more powerful, adding on to today's features like DVR with new features, such as home networking and whole-home DVRs that can record six or more channels and send the recorded programs to any TV in the home.

Cable companies hate paying for expensive set-top boxes, but they like the concept of having a piece of gear in your home that controls much of your TV viewing (and maybe even your Internet surfing and phone calls, too!). So don't be surprised to see your cable company still pushing set-top boxes your way, even long after CableCARD 2.0 hits the streets.

Because of its limitations (no program guide, no VoD), users haven't overwhelmingly accepted the CableCARD 1.0 standard. And CableCARDs aren't available on all HDTV sets — particularly on the popular flat-panel LCD and plasma TVs. (As we write, only 19 of the 63 HDTV-compatible flat panels for sale on Crutchfield.com support CableCARD.) Nor can you yet find CableCARDs in standalone DVRs (like the TiVo or in a Media Center PC; see Chapter 16 for more on these PCs). Most manufacturers appear to be waiting to see what CableCARD 2.0 will look like when it appears on the market, probably some time in 2007.

Although CableCARD isn't yet all that it can be, it's still a pretty handy way of getting your HDTV (and other digital cable programming) while saving a few bucks and avoiding yet another box connected to your HDTV system. So if you can live without an on-screen guide and VoD, we recommend you at least check it out. The biggest issue we have is the lack of DVR support — we gotta have DVR in our home theaters, and right now the only way to get DVR recordings of all of your HDTV programming in a cable system is to use the cable company's set-top box and DVR service. The new Series3 TiVo DVR, discussed in Chapter 12, will solve this problem.

Chapter 9

Rocket Science

In This Chapter

- Figuring out the pieces and parts
- Sifting through the offers

Well, okay, rockets *did* have something to do with putting all those communication satellites up there. But that's pretty much where the science ends and the good news begins. Satellite service providers are not only a great alternative to cable for regular TV programming, they're also a primo source for HDTV. The satellite companies were among the first to offer HDTV — using HDTV as a serious competitive advantage to counteract cable companies' digital cable offerings, such as video on demand (VoD).

With its very small (usually 18- to 24-inch) dishes, *direct broadcast satellite* (DBS) allows all but the most remote viewer to receive HDTV economically. Even author Danny can receive satellite HDTV on his island two miles out at sea.

But you don't need to be out of range of cable or terrestrial broadcast systems to consider satellite HDTV. Satellite systems have many advantages, and many folks right smack dab in the middle of town opt for satellite.

Planning a Satellite TV System

Getting satellite TV for your HDTV isn't all that difficult, but it is a bit more complicated than cable TV (where most of the sophisticated equipment belongs to the cable company, so the company handles all the maintenance and installation).

The first step towards satellite involves picking a provider (imagine that, a choice instead of a monopoly!). After you pick a provider, you can choose the appropriate equipment — a satellite dish and a receiver.

Service

You must subscribe to a provider if you want HDTV channels from a satellite. In the United States, two TV service providers have significant HDTV programming:

- DIRECTV (www.directv.com)
- DISH Network (www.dishnetwork.com) owned by EchoStar

A few years ago, a third provider — Voom Networks — made a big splash by offering a third option for HDTV (and regular TV) via satellite. Unfortunately, Voom didn't last long as an independent satellite TV provider. But the good news is that Voom's 15 unique HD channels are still available through DISH Network. Check out www.voom.tv for more info.

The second half of this chapter guides you to select a provider for satellite HDTV.

Equipment

For HDTV (or any TV, for that matter) by satellite, you need equipment to use the satellite signal: a receiver, an antenna, a TV, and cables to connect them all.

Want more information than that? Can do!

Receiver

Satellite HDTV requires an HDTV *receiver* (sometimes called a *tuner* or *set-top box*) that works with your DBS provider.

If you already have a DIRECTV or DISH Network receiver, it probably *doesn't* receive HDTV. Check the Web sites for DIRECTV (www.directv.com) and DISH Network (www.dishnetwork.com) for the latest HDTV receivers.

A big reason why you would need to upgrade an existing satellite receiver to a newer HD-capable model is the fact that most satellite TV providers have switched or are in the midst of switching the *compression* system they use for transmitting TV signals over the satellite link. Traditionally, satellite TV providers have used *MPEG-2* compression, but as they add more and more HDTV channels (and local TV channels, as well), they've begun to use the more efficient *MPEG-4* system. This system is more efficient because it can fit more channels in the same electromagnetic spectrum. All the new HD-capable satellite receivers that DISH and DIRECTV sell support MPEG-4.

Except for special HDTV connections, an HDTV satellite receiver usually looks and works just like a typical satellite TV receiver. Most HDTV satellite receivers include both

- A receiver for satellite-based HDTV channels
- A *terrestrial* broadcast HDTV tuner so you can pick up local HDTV channels that aren't on your satellite lineup

Local HDTV broadcasts are transmitted *digitally.* Expect any HDTV satellite system to get a fine picture from local HDTV broadcast stations unless the local signal has to travel so far that it's just too weak when you get it. Chapter 7 gives you the lowdown on receiving over-the-air broadcasts of local HDTV channels.

A few TV sets have *integrated* DBS receivers. Each receiver is for only one DBS provider, so you need to decide on a provider before buying an HDTV set with an integrated receiver.

If you want a *digital video recorder* (DVR) in your HDTV satellite system, it should be part of the satellite receiver. Later in this chapter, we tell you about DVRs that are available for DIRECTV and DISH Network. Chapter 12 has the whole DVR story.

Antennas

Even if you have a DBS dish, you might need a new dish for HDTV. DBS providers have launched new satellites to transmit HDTV (and local channels and other services) because normal DBS satellites are pretty much full to the brim with regular TV programming.

The DBS company (or the installer you use for your HDTV system install) usually installs *two* HDTV antennas for you:

- A small broadcast antenna that receives local HDTV channels
- The special HDTV dish for your satellite channels

Why might I need a new dish?

HDTV signals are transmitted in their own *orbital slots* (from different satellite locations, in other words). For example, DISH Network broadcasts HDTV from the 110-degree slot, which many older DISH Network dishes (that's a mouthful!) can't tune in. In those cases, you need a new satellite dish that incorporates low-noise blockers (LNBs) that can tune in these new slots before you can receive the HDTV signal.

If you already have DBS service, check whether you need a *new* satellite dish to get HDTV satellite channels from the same DBS provider:

- **If you have DIRECTV service,** look for a DIRECTV logo on Channel 99. If you see the logo, you're set for HDTV from DIRECTV. Otherwise, you need to install a new dish (or in a few cases, purchase an upgrade kit for your existing dish).
- **With DISH Network,** try tuning to channel 9900. If you receive this channel, you have the right dish; otherwise you need to install a new dish.

If you're looking to get local channels over your dish, these tests might not work. The best way to ensure your existing dish works is to call your provider and ask.

If you need a new satellite dish, let a pro install it. Running cables and aiming the dish can be pains in the hind end. There's no real advantage of doing it yourself, especially if you can get *free* installation in a package when you subscribe.

HDTV-ready TV

Your new HDTV-ready TV connects to the satellite HDTV receiver the same way it connects to a similar cable box. Chapter 1 and Chapter 2 guide you to the right HDTV-ready TV for your home.

If you don't have an HDTV yet, you can use an HDTV satellite receiver with your existing TV. You won't see HDTV-quality pictures with such a setup, but it works. You might do something like this, for example, if you're waiting for a specific HDTV but you're ready to install or upgrade your satellite TV system now.

Cabling

An HDTV satellite system needs three kinds of connections:

- **Satellite dish to satellite receiver:** When the signal arrives from outer space, it has to get to the receiver before you can see it. No rocket science there.

  If you already have a DBS system, you probably can use the old cable to connect your new HDTV dish to your new HDTV satellite receiver. But let a pro install and aim your dish.
- **Local antenna to satellite receiver:** Chapter 7 shows how to receive HDTV broadcast stations on HDTV devices.
- **Satellite receiver to TV:** Chapter 3 covers basic video connections to an HDTV-ready TV.

Availability

Most U.S. homes can receive HDTV by satellite. Here's how to make sure *you* can receive it:

- **Go surfing:** The best way to find big problems or special needs is to check each provider's Web site. Zip-code-based information portals can tell you what's needed for your area.
- **Check your dwelling rules:** If you are in an apartment, condo, or a homeowner's association, check your local rules about dish placement. In the United States, you must be allowed to install a dish (that's a federal regulation), but the dish must be located in a space that only you have access to, such as a private patio, and not on common space, such as the rooftop. If your permitted placement doesn't have a southwestern sky view, you need to negotiate with your landlord or condo/homeowner's association about installing the dish elsewhere (for example, the roof). Check out the FCC's Web site (`www.fcc.gov/mb/facts/otard.html`) for the details on this ruling.
- **Check your latitude:** If you're in Alaska or Hawaii, you might need a larger dish.

  For example, DIRECTV advises its customers that subscribers in Anchorage or Fairbanks might have to install an 8-foot dish to get their signals. (You can probably get by with a 4-foot dish in Juneau.) If you live to the north and west of these locations, signal strength continues to drop off, but we're guessing if you live that far north of the Lower 48, you probably know this as a matter of course.

Special international channels, such as Chinese- or Spanish-language channels, might come from other satellites. They require a multisatellite dish and visibility to all those satellites.

Choosing a DBS Provider

Your home probably can receive both satellite-HDTV providers. If you can't decide by features and price (most folks who live in medium-sized and larger cities can decide that way), the trick becomes sorting through the offers *du jour.*

At the time this was written, DISH Network was offering a limited-time, free-equipment-and-installation deal to drum up business, whereas DIRECTV was packaging a free HDTV receiver or DVR upgrade for existing customers and a rebate up to $100 plus a free portable DVD player for new subscribers.

Special deals come and go, so you need to check with both vendors for their packages when you're ready to buy.

Programming

All the HDTV satellite providers have the common channels you expect in any cable or satellite system. But actual HDTV channels vary from provider to provider.

DISH Network channels

DISH Network offers 30 network HDTV channels, including

- **Six HD movie channels,** including HBO and Showtime in HD
- **Five HD sports channels** including ESPN and ESPN2 in HD, as well as NFL Network Rush HD and Worldsport HD (great for any soccer fans feeling left out in the United States)
- **An HDTV pay-per-view channel,** featuring recent movies and some IMAX films
- **HDNet,** which carries live sporting events, and a variety of dramas, documentaries, and other programming
- **HDNet Movies,** which shows (you guessed it!) movies from Warner Brothers, Sony Pictures Television, and independent sources

A big part of DISH's HD offerings are the 15 unique channels sourced from Voom — you can read more about these in the sidebar titled "Va Va Va Voom!"

In about 50 percent of the United States, DISH subscribers can pick up their local broadcast channels via the satellite dish (including the high-definition programming carried over these local channels). The other half of the country needs an over-the-air antenna to pick up these channels (an ATSC tuner is built into high-definition receivers for DISH).

All but the premium movie channels (HBO, Showtime, and Starz) are included in DISH's $49.95 per month plan, which also includes 80 other standard-definition channels.

DIRECTV channels

DIRECTV offers 13 channels of HDTV broadcasts of popular networks (including HBO, Showtime, TNT, Discovery, and ESPN), plus a pay-per-view HDTV offering.

Both DISH and DIRECTV offer Universal HD, which means you can get Battlestar Galactica in high definition (which Pat's local cable company doesn't offer him — boo!). We just thought you'd want to know!

Va Va Va Voom!

Back in 2004, Voom was a new satellite TV provider that was launched (literally as well as in the more figurative "corporate launch" sense) to compete with DIRECTV and DISH Network in the North American market. Voom was really interesting to HDTV owners as it had nearly 40 high-definition channels — many more than any other TV source at the time. Well, unfortunately for those HDTV owners and also for the folks who invested many millions (if not billions) in launching the satellites, Voom didn't make it in the marketplace, and shut down in 2005.

But that wasn't the end of Voom (which is a subsidiary of the large cable company Cablevision). Voom lives on as a suite of HDTV channels that are packaged for other television providers to include in their own HDTV channel lineups. And if you're a DISH Network customer, you can add this package of 15 unique and high-definition channels through your satellite dish and receiver.

Voom's offerings go beyond the typical HBO, ESPN, and network offerings and consist of unique HD-only channels. These include such unusual offerings as

- **Equator HD,** which broadcasts "intriguing and visually stunning sights and sounds that capture the world's most unique people and places" — that is, documentaries and other visually appealing fare.
- **Rush HD,** which features HD broadcasts of extreme sports. If you want to see someone ride a skateboard off a cliff for some extreme base-jumping action, tune in here.
- Danny's favorite is **Kung Fu HD,** which features all the martial arts action you can shake a fighting stick at, and all of it in high definition.

With Voom, you can also watch Westerns on the Gunslinger channel, plus classic or epic movies (or epic classics, no doubt).

Today, Voom is available only on DISH Network, but the company plans to offer these channels internationally and through cable networks, as well. If you're interested in keeping up on the latest Voom news, you can sign up on its Web site (`www.voom.tv`) for the latest news and announcements via e-mail.

Eight channels (ESPN and ESPN2, Universal HD, TNT HD, Discover HD Theater, HDNet, and HDNet Movies) are included in a $9.95 a month package on top of your standard-definition lineup fees.

The big HD benefit of DIRECTV is in the sporting realm, where you can purchase packages for NBA Games, NFL Sunday Ticket (see all the games in HD!), and (nearest and dearest to our hearts) Mega March Madness in HD. Go Duke!

Like DISH, DIRECTV offers local broadcast channels in HD over the satellite. In 36 markets (as we write in mid 2006), you can pick up CBS, NBC, ABC, and Fox in HD for no additional fee over the standard-definition service.

DIRECTV is in the midst of a huge expansion in satellite capacities and is also moving to the MPEG-4 system for all HD broadcasts. By the end of 2007, DIRECTV expects to be able to offer over 150 channels of high-definition programming nationwide and will be able to tap into 1,500 local HDTV broadcasts. (You can receive only those broadcasts local to you, not all 1,500.)

Digital video recorder (DVR)

Digital video recorders let you pause, fast-forward, rewind, replay, and *time-shift* (record and watch at your convenience) your favorite programming. (We cover DVRs in depth in Chapter 12.) A DVR is more powerful and easier to use with an HDTV satellite system if the DVR is *included* in the satellite receiver.

Both satellite providers offer an option to purchase an integrated high-definition receiver/DVR device:

- **DIRECTV:** DIRECTV has a new multiroom receiver/DVR combo called DIRECTV Plus HD DVR. This super box can tune in or record two high-definition or standard-definition signals at once. It can also do all the magic DVR stuff we talk about in Chapter 12, like pause or rewind live TV (it caches up to 90 minutes of programming), and by the end of 2006, it will include the capability to record over-the-air digital (standard- or high-definition) broadcasts. Plus HD DVR also includes on-screen caller ID (because you need to have a phone line plugged into the receiver for accounting and service authorization purposes), and features one-button recording setup, and the ability to record an entire season of a show, even if show times change. Figure 9-1 shows Plus HD DVR, which costs $299 for new subscribers. However, you can often find discounts and rebates in effect — as we write, it's $100 off. Check the Web site at www.directv.com to see what's current when you're shopping.

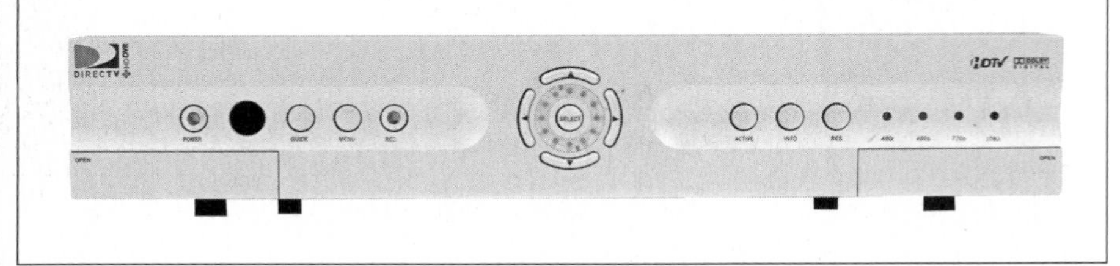

Figure 9-1: Going HD and DVR with DIRECTV Plus HD DVR.

Until recently, DIRECTV offered a TiVo-powered HD DVR receiver, but the new model is its own system, without the TiVo software. We mention this only because there's a sizeable group of folks out there who are TiVo fanatics, and they might be disappointed by this. For them, the only TiVo solution for HDTV is the new cable and over-the-air TiVo Series3, described in Chapter 12.

- **DISH Network:** Like DIRECTV, DISH Network has it own fancy-pants HD DVR receiver, the ViP622. This is a two-room receiver, meaning it can display DISH programming both on the HDTV to which it is attached, and also on a second TV in another room (or even in the same room — DISH won't be sending any Matrix-like agents into your home to make sure that second TV is elsewhere in the house). The second TV *isn't* going to get an HDTV signal; it's just standard definition, running over a standard coaxial cable and plugged into the back of your second TV. However, the system can *down-convert* HD channels to standard definition, so you can watch your HD-only programming on the second TV at a lower resolution.

  Like the DIRECTV DVR, the ViP622 has two tuners (plus two ATSC tuners for over-the-air broadcasts), so you can record two things at once, or watch one live while recording another, or even record two while watching a third *already recorded* show. It also includes on-screen caller ID and all the DVR features that make people happy. With a service contract, the ViP622 (shown in Figure 9-2) is $199.95.

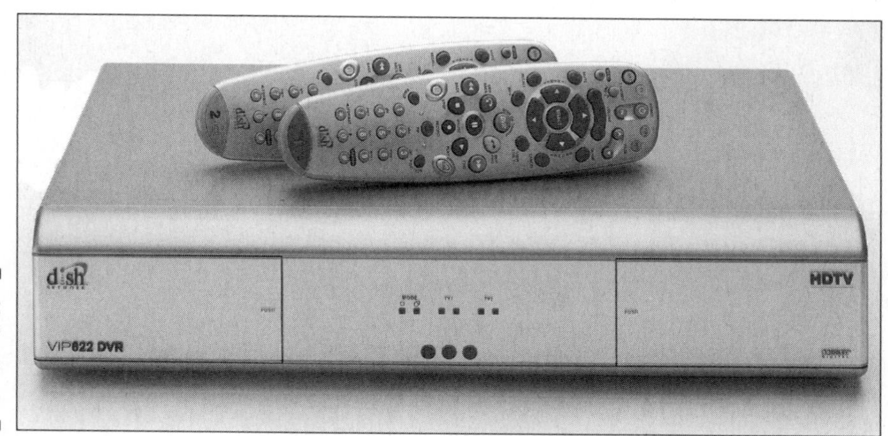

Figure 9-2: Being a VIP with the ViP622.

Internet and HDTV with one dish

It might be the industry's best-kept secret, but you can get both high-speed Internet (via DIRECWAY) and DIRECTV's HDTV service from a single installed dish. The dual dish comes with *three* connecting cables:

- A cable to the HDTV satellite receiver.
- Input and output cables for Internet service. The input and output cables connect into a high-speed modem, which then connects into the computer.

You can share DIRECWAY Internet service with more than one computer in your home by either connecting computers to your DIRECWAY service through a home network or purchasing a DIRECWAY modem that can connect more than one computer.

DIRECWAY is a godsend for anyone living on an island or remote mountaintop, where you'd never get any other Internet access. (Ask Danny — he knows.)

Compared to other Internet services, DIRECWAY has a couple of drawbacks:

- DIRECWAY is a fast Internet service, but it adds a delay in all your Internet activities. Its signals must travel 23,000 miles up to satellites and then the same distance back down. That adds a half-second to every round trip. The delay isn't significant for e-mail, text messaging, and Web-surfing, but you'd notice it if you play games or use voice over the Internet.
- DIRECTV and DIRECWAY haven't been offered together as a special deal. (Cable and telephone companies often give you a discount if they sell you TV and Internet service together.) But that can change. If you're considering both DIRECTV and DIRECWAY, look for combination deals that might have been created since this book was published.

If you intend to install DIRECWAY, ask for the dual dish at a DIRECTV retailer. As we write, DISH Network has announced a similar service through a new partnership with a company called WildBlue (which also helps DIRECTV provide its service). We haven't seen all the details and requirements yet, but we anticipate that they're similar.

The bottom line is that if you live somewhere where you can't get cable or DSL, either of the satellite TV providers can help you get online.

Chapter 10

Internet-Based HDTV

In This Chapter

- Surfing for your Surfer movie, dude
- Putting Internet video onto your HDTV
- Taking your HDTV with you on the road
- Finding the best way to watch your HDTV remotely

If "all the world is a stage," the Internet must be the whole dang theater! You can find anything and everything on the Internet — you just have to know where to look. That's where we come in. We help you figure out those nooks and crannies of online HDTV and other audio and video content for your HDTV theater experience.

As we write this, the world is in an Age of Video Discovery of sorts, with Google, Yahoo!, iTunes, YouTube, and others all trying to become your preferred path towards getting video content. In the end, we think, they'll all offer the same fare for about the same price. So where you shop or go for free content comes down to a matter of preference, ease of use, and other features.

In this chapter, we expose you to what content is available and in what online venues, and we cover how to get it into your home. Get ready. Prepare to download content. Click.

Understanding Internet Content

Internet video content ranges from grainy live webcams to full-length, high-definition, director's cut Hollywood movies, and lots of gradations of quality in between.

One of the most rapidly growing areas of the Internet is amateur video — non-copyrighted video shot on vacations, around the house, and in those damp basement studios. Online destinations like YouTube shot from nothing to mega-success because they allow people to easily load their home videos onto the Internet. (See the "How to prepare video for YouTube" sidebar if you're interested in finding out how to upload your video.)

Streaming versus downloading

You can watch video that is stored on the Internet in two ways:

- **Streaming video:** With *streaming* video, your video-reading program starts to download the video from the source site into a buffer in memory or on your hard drive. The program monitors the download speed of the content and starts the video when enough content is downloaded to begin playing the video and continue through to the end without pause.
- **Downloaded video:** With downloaded video, the entire program is downloaded prior to showing. Oftentimes, you can allow content to download to your system in the background while you're watching something else or when you're asleep. Some systems, like MovieBeam (`www.moviebeam.com`), download lots of movies — 100 in MovieBeam's case — so you can start watching the first video right away without having to wait for all the movies to download.

With the dramatically increasing size and decreasing cost of hard drive space, we think it will become quite commonplace to have lots of movies stored around your house, protected by e-commerce and DRM (digital rights management) schemes, which make the movie available to you only when you buy it. You won't need to actively be downloading movies from the Internet — it will be something that continually happens in the background of your high-speed Internet connection.

Most content found on the Internet is *not* full high-definition resolution because of how long it takes to download. As broadband speeds increase from 1 Mbps to 10 Mbps to 100 Mbps, more HD content will appear. Also, as local storage space increases within devices in the home, more content can be preloaded for later viewing — this too will boost demand for HD movie downloads.

Understanding containers and codecs

Digital media files are often described by the *container format,* which is a special kind of computer file format designed to contain (no duh, huh?) media. The container file defines the type of media within the file and is used by your computer or media player to figure out what kind of media it is dealing with. (We discuss media players in the section "Understanding Your Hardware Requirements.")

Each container file (examples include .asp files for Windows Media and .mov files for QuickTime files) holds media data that has been encoded digitally using *codecs*. Codec, which is short for encoder/decoder, can mean either compressor-decompressor or encoder-decoder — either way it's a standardized way of taking an analog audio or video signal and turning it into a computer-readable digital file. For example, a Windows Media container typically uses Microsoft's .wma and .wmv audio and video codecs.

The combination of containers and the codecs within the containers are the "secret sauce" that lets a computer or media player figure out what kind of media you have and how to turn it back into video you can see and music you can hear.

Understanding file formats online

Internet-based video is encoded in different formats (called codecs), and you need to be able to decode that content to view it. The encoding process basically takes the content and optimizes it for viewing in different instances. You wouldn't want to watch a full HDTV-encoded movie on your cell phone, for instance — movies broadcast over the cellular networks are encoded in formats specifically designed for viewing on a 2-inch screen.

The most popular encoding formats are

- **Windows Media:** This is Microsoft's format for video and audio content (.wmv and .wma, respectively). You can download Windows Media Player from www.microsoft.com to handle these formats on your PC.
- **MPEG:** MPEG stands for Motion Picture Experts Group, which is a standards organization that develops video codecs used throughout the world. There's more than one type of MPEG (in fact, the popular MP3 music file format is a variant of MPEG), but the most common are
  - **MPEG-2:** This is used for many digital television systems. Files encoded in the MPEG-2 format typically have the .mpg file extension.
  - **MPEG-4:** This is a newer and more efficient codec. It uses less storage space to hold the same quantity and quality of video as MPEG-2. MPEG-4 is also known by the catch name H.264 and AVC — advanced video codec. Files encoded with MPEG-4 will often have the .mp4 file extension.

  On the computer, most media players (like Windows Media Player, Real Player, and QuickTime) can handle MPEG.

- **QuickTime:** The Apple media format is QuickTime. QuickTime is a very versatile media format that can include many different encoding systems (like MPEG) as well as the Apple proprietary codecs. QuickTime players are available for both Windows and Apple computers from `www.apple.com`. QuickTime files typically have the `.mov` file extension.
- **DivX:** A very popular video-encoding system used for Web-based video is DivX. The DivX codec itself actually uses a variant of H.264 MPEG-4, but with DivX's own proprietary version of that codec. You can find a PC (and Mac and Linux) player for DivX content at `www.divx.com`, and many DVD players can play DVDs that are encoded with this system. DivX is a commercial system, which means you might have to pay for software to create your own DivX content. However, a very similar open-source codec called XviD (`www.xvid.org`) has been developed to provide an alternative. DivX files typically have a `.divx` file extension, but the DivX codec can be found within other file types such as Windows Media.

High-definition beauty, Hollywood style

The great thing about movie stars is that because they make so much money, you don't mind being catty about them. Everyone loves it when they trip, divorce, or get a DUI because, well, it's just human nature to want to see those who are successful, fall (sometimes literally).

So it comes as no surprise that someone would conjure up an annual scale to rate which actors and actresses look the best and worst on HDTV! That's right, a veritable "Who's the best looking and ugliest, up close and personal!" or at least who appears that way when shown in high definition. High-definition TV is very unforgiving because every detail comes through very clearly, and leave it to Phillip Swann, president of TVPredictions.com, to turn rating HDTV presence into an annual rite of passage for Hollywood.

Since 2004, he has been naming his "Best and Worst HDTV" list. As we write, his 2006 "HD Horribles" list has Rosie O'Donnell at the top, and his "HD Honeys" list named youthful Scarlett Johansson as "worth a thousand looks." Howard Stern, Teri Hatcher, and Britney Spears follow Rosie as needing some high-def help, but Beyonce Knowles, Josh Holloway, and Eva Longoria are on the "Honeys" list, along with others. You can see the whole list here: `http://www.tvpredictions.com/fall2006hd092706.htm`.

Being *Lost* fans, we had to chuckle at his write-up on Honey Number Three, *Lost* star Evangeline Lily: "In high-def, the luscious *Lost* star makes being marooned look like a vacation. No wonder Sawyer and Jack seem to have such little interest in getting off the island." Maybe the Professor did have something for Mary Ann, huh?

So as politically incorrect as these lists are, it's no worse than Danny's *People* subscription that he doesn't want you knowing about. Oops!

One thing to remember about encoding systems is that any particular codec can vary in its quality and performance. That is to say, someone can encode video and audio content using more or less compression within a single codec. For example, you can find very low *bit rate* Windows Media movies (the postage stamp videos embedded in Web pages) or very high-quality HDTV-ready Windows Media movies. You can find a large degree of latitude in how a particular codec is used to encode a movie or video program. There's always a trade-off between size (how much hard drive space a video takes up and how long it takes to download) and quality (the higher the bit rate, the higher the quality).

Finding a Provider Online

The options for finding video online have blossomed in recent years. Where video used to be an oddity, now it's everywhere including on Web sites, in blogs, and even in e-mails.

Some sites used to offer free video and some sites required payment for video, but those distinctions have become lost as most "free" video site owners have concluded deals with the owners of copyrighted videos to sell those videos on their sites. So it's just a matter of time before most video content is available on most video distribution sites. At that point, sites will become specialized and built into other portals (or Web destinations aggregating content from multiple sources) specific for certain topics, like rodeo videos or space adventures.

So you should view and buy video from the site you're most comfortable with and find other reasons to visit. Maybe it's where you get your news (Yahoo!), where you search (Google), or where you buy your music (iTunes).

The following sections look at some popular download destinations.

Apple iTunes Store

The most popular place on the Internet for legal audio downloads is the Apple iTunes Store (`www.apple.com/itunes`). Apple recently changed the name of the store from "iTunes Music Store" to just plain old "iTunes Store" for a reason: Apple has branched out way beyond plain old audio and has begun selling tons (millions!) of videos. Here's what you need to know about iTunes:

- To access the iTunes Store, you need to use Apple's iTunes application (available for free download for Mac OS X and Windows).
- The videos for sale as we write are optimized for display on a PC or an iPod screen — not for your big-screen HDTV. We think this will change shortly, based upon Apple's forthcoming iTV device (discussed later, in sidebar titled "Coming soon: The Apple iTV"), and that Apple will soon begin offering HDTV downloads.

As it stands today, you can purchase and download hundreds of TV shows and a growing number of full-length movies.

YouTube

YouTube (`www.youtube.com`) is a sensation that has taken the Internet by storm over the past year or two — so much so that as we write, Google has just plunked down over 1.5 billion (we like to draw that out like Dr. Evil; billlllllllllllllllllllllllion) dollars to purchase YouTube. YouTube videos are typically not all that exceptional in their quality; they're not going to replace your Blu-ray high-definition discs any time soon. But what they lack in ultimate picture quality, they more than make up in fun.

YouTube is chock full of fun, funny, and just plain interesting *user-generated* content — homemade movies that people put together and post out in the wilds of the Internet for any and all to download. A lot of this stuff is simply short clips from regular TV shows (if you're a fan of the Colbert Report, you can probably find your favorite clips on YouTube), but increasingly the best stuff is made by millions of budding Scorseses out to make the next great independent movie. As YouTube absorbs its billllllllllllllllllllions of dollars and as HD camcorders and editing software make it into the hands of many folks (see Chapter 15 for info on these camcorders), we expect that you'll start seeing some really compelling HD content on YouTube.

Google Video

Google didn't just get into the video business on the day it purchased YouTube. In fact, Google has a huge number of videos available for download from its video service (found at `http://video.google.com`). These range from the user-generated videos that YouTube is famous for to downloadable commercial videos that you pay for. For example, if you're a fan of our fellow

Duke graduate Charlie Rose, but you forgot to DVR that 12:30 a.m. showing, just pop onto Google Video and download it for $0.99!

Movielink

Here's where you start bumping into the kind of content that really can compete with DVD rentals and pay-per-view (PPV) movies from cable and satellite providers. Movielink is a joint venture of several major movie studios designed to provide easy downloads of full-length movies to your PC hard drive. Movielink offers both rentals and purchases. When you rent, the movies are downloaded to your PC and expire after a fixed period of time. When you purchase, you pay a bit more, but you can keep the movie forever and even burn it to DVD and play it back in any standard DVD player.

So far, Movielink doesn't offer high-definition versions of its movies, but this is on the agenda for the Movielink folks, and we expect it to be available soon. Check it out at `www.movielink.com`.

You can find movie download services similar to Movielink at `www.cinemanow.com` and `www.vongo.com`.

How to prepare video for YouTube

Getting your content onto places like YouTube, Google Video, or Yahoo! Video is easy. Software packages on the market like Movavi's VideoSuite ($59.99, `www.movavi.com`) give you the means. With VideoSuite, you just

1. **Shoot your movie with a digital camcorder, camera, webcam, or cell phone.**
2. **Save the video to your hard drive or burn it to DVD.**
3. **Launch Movavi VideoSuite.**
4. **Click the Convert Video/DVD option in the window that appears.**
5. **Add your video or DVD file to the list of open files by clicking the + Video or + DVD button on the toolbar and navigating to the file in the file dialog box that opens. Just double-click the filename when you find it.**
6. **From the Output Format drop-down list, select AVI.**
7. **From the AVI Pre-Sets drop-down list, select "MPEG4 XviD for YouTube" or "MPEG4 DivX for YouTube."**
8. **Click the Convert button.**
9. **Save the new file.**
10. **Upload to YouTube.**

Pretty simple, eh? (It's even more fun with one of those sexy HDTV camcorders we talk about in Chapter 15.)

Amazon.com Unbox Video

Not satisfied with being the largest online seller of DVD movies in the world, Amazon.com has launched a new service designed to make it a major player in the movie and TV download business, as well. Called Unbox (www.amazon.com/unbox), this service offers hundreds of TV shows and feature-length movies for download. You can either rent or purchase (and then save to DVD) Unbox movies and shows, but unfortunately you can't burn your own *movie* DVDs with Unbox yet. That is to say, you can store movies on a recordable DVD, but you have to play them on your PC and not on a regular DVD player.

Unbox is still relatively new as we write, but given the large amount of content already available, we expect it to grow quickly.

Akimbo

The preceding Internet video services are very PC-centric. You access them via a TV and a Web browser, and the movies are downloaded to your PC. To display them on your HDTV, you need to have a PC connected directly to the HDTV.

An exception to this model comes from the folks at Akimbo (www.akimbo.com). Akimbo has a very unique model (which many other Internet TV providers are starting to adopt) that combines Internet TV programming with special hardware that connects directly to your HDTV. Akimbo's service includes over 100 channels of Internet-delivered movies and shows, including movies from Movielink, Major League Baseball highlights and classic games, BBC programming, and even a special National Geographic Channel. Akimbo is priced much like a "regular" TV service, with a number of channels included for a monthly service fee, and other channels available as premium or pay-per-view services for extra money.

You can get Akimbo content to your HDTV in two ways:

- **A Media Center Edition PC that's connected to your HDTV:** You can use it to subscribe to Akimbo or you can buy Akimbo's new set-top box built by RCA. The RCA box connects to your home network and to your HDTV, and it lets you access all of the Akimbo service via an interface on your TV screen. See Chapter 16 for more about Media Center Edition PC.
- **Satellite TV service through AT&T (AT&T resells DISH Network):** You'll soon be able to access Akimbo directly through your satellite receiver. AT&T's new Homezone receiver will include an Akimbo client that works just like the standalone RCA box, and you can even include the Akimbo service on your monthly AT&T bill. (You can pay for video, Internet TV, broadband, and telephone all on one bill — think of all the money you'll save on stamps!). See Chapter 9 for more on DISH.

What's peer-to-peer all about?

If you remember the early days of digital music on the Internet (pre-iTunes, that is), you probably recall services like Napster (the original Napster, not the current music store that has the same name). These were *peer-to-peer* file-sharing services that let people all around the world share digital media. Peer-to-peer systems don't rely upon centralized servers that send individual copies of a file to each user — instead, files are distributed on users' computers and are made available for download directly to other users or *peers*.

Peer-to-peer can be used legitimately, and many folks are looking into how this might make video downloads faster and more inexpensive, but a lot of stuff available on peer-to-peer networks (using programs and protocols such as BitTorrent, at `www.bittorrent.com`) is copyrighted and not really supposed to be available for free downloads. If you download a lot of content from peer-to-peer networks and it's copyrighted, you could be placing yourself in legal jeopardy. Having said that, if you choose to explore this world, you can often find pretty much any program or movie you're interested in, and often in high definition . . . just make sure it's legal!

Understanding Your Hardware Requirements

All the content on the Internet won't do you any good if you can't get access to it. You need the following at a minimum to download and play video content from the Internet:

- **Broadband Internet connection:** A cable modem, a DSL line, a satellite downlink, a wireless connection . . . any of these can do the job; the faster the better.

  The cable and telephone companies are in the midst of a massive battle, and you're going to be the big winner. One tool to win customers is sheer bandwidth. Cable modem and telephone company DSL download and upload speeds have jumped from early 384/128 Kbps levels to almost 100 Mbps in some areas. Verizon's FiOS services (`www.verizon.com`), for instance, routinely provide users with 50 Mbps download speeds — that's insane. At that speed, you can download a 2 hour HDTV movie in just 24 minutes. Wow. (We're jealous because we don't live in areas where the service is available.) So get the fastest speed you can for video downloads.
- **Player hardware:** This could be a set-top box, PC, gaming platform, or any device that can connect to the Internet and play video content. Some devices are specifically for downloading video from the Internet, like the RCA Akimbo set-top box we mention earlier. Some devices that connect to the Internet can also play downloaded content, such as a Media Center PC.
- **HD display:** You need a high-definition screen to watch it on. We think you have that part covered!

Modding your original Xbox

The new Xbox 360 makes for a great Internet video extender from your PC, but you don't need to drop $500 on one to get this feature. The original Xbox can also serve as a Media Center extender, and Microsoft sells a hardware/software package that does just this. See `www.xbox.com` for details.

An even more full-featured and powerful way of using an original Xbox as an Internet TV source for your HDTV is to install an open-source program called Xbox Media Center on your Xbox. This program turns the Xbox into what it really is at heart: a powerful, graphics-oriented computer that can put pretty pictures up on your TV screen. You can find details about it at `www.xboxmediacenter.com/wiki`.

Xbox Media Center is a free download, and it supports all sorts of media types over your home network. But there's a catch (sorry!): You need to have a *modded* Xbox. Modded Xboxes have a special chip (called a *mod chip*) installed inside the unit that bypasses the original Microsoft *BIOS* (the initial software instructions that load when the Xbox is started) to allow you to do things that Microsoft never intended you to do with an Xbox.

If you install a mod chip in your Xbox, you do two things:

- You kind of peeve Microsoft. (Microsoft is constantly trying to shut down the folks who build the mod chips.) And while it might go without saying, we're going to say it: If you mod your Xbox, you'll no longer be covered by Microsoft's warranty. This might not be a huge concern if your Xbox is older and out of warranty already, but if it's still covered, think twice!
- You lose your ability to use Xbox Live for online gaming. So this isn't for the faint of heart, but it is possible. We can't tell you where or how to get a mod chip, but a simple Google search can provide you with a lot of detail.

You need all three components for Internet video on your HDTV.

The hardest part of this whole Internet TV thing is making the connection between the Internet and the TV. You have three main options here:

- **Connect a PC directly to your HDTV.** This is the easiest solution in many ways because the HDTV just becomes a big PC monitor. With a Media Center Edition PC or a home-theater PC using a different system (see Chapter 16), you can simply fire up the PC, switch your HDTV to the input that your PC is connected to, and go to town! Of course, this is a rather pricey solution because most Media Center Edition and home theater PCs cost $1,000 or more. But if you want Internet content on your HDTV and you want it now, and if you want it to be easy, this is the way to go.

- **Extend your PC through a gaming console to elsewhere in the home.** The Microsoft Xbox 360 is the perfect example here. It comes out of the box with a Media Center extender client that lets you easily access videos and other media stored on a Windows PC on your home network. Sony hasn't specifically announced a similar functionality for the forthcoming PlayStation 3, but we'd be surprised not to see something like this as well. And these consoles have Internet capabilities of their own, so online providers can send video directly to them rather than storing it on a PC and then sending to the game console. If you're thinking about buying one of these new gaming consoles (see Chapter 14), this method can be great for getting Internet TV to your HDTV cheaply.
- **Connect your PC over the home network to a digital media adapter or media player.** Then you can convert the video to a format that your HDTV can display. Dozens of these devices are on the market today (we talk about a few coming up), and they're typically quite inexpensive (usually under $150). The biggest issue here is that not too many of these devices can support high-definition video. They're limited by hardware and network limitations to standard-definition video, and often this video isn't even all that great. Having said that, a lot of work is being done by all the major home networking manufacturers to turn these devices (or their next generation successors) into true HDTV sources.

Instead of a digital media adapter or media player, you can connect your PC directly over the Internet to a source of Internet video.

Coming soon: The Apple iTV

The folks at Apple have an answer to the media adapter question, and it's coming soon. Apple pre-announced a device codenamed *iTV* in late 2006 and will begin shipping it sometime in 2007. We don't know a lot about the iTV yet, but we have a few facts:

- It's expected to cost about $300.
- It looks a lot like the Mac Mini: It's tiny, sleek, and white (any color you want, as long as it's white).
- It's got an HDMI output port, so it's bound to have *some* sort of HDTV support — though the initial statements by Apple are promising only DVD-quality video.
- It's designed to work with the Apple iTunes Store.

Apple has begun really focusing on video on the iTunes Store, though to date the content has all been low-resolution video designed to be played on an iPod or on a PC using iTunes software. Speculation (and that's all we have to go on) is that iTV will take this to the next level (to high-definition heights) and will offer HDTV owners the high-def content they so strongly desire. There's a bit of a wait-and-see here, but we're very intrigued. Apple didn't become the digital music powerhouse it is by messing up big new product introductions, and we expect that the ease of use and seamless integration that iTunes and iPod offer will be replicated with the iTV.

Using "sneaker-net" for your Internet content

Throughout this chapter, we discuss a bunch of networked systems that enable you to get Internet and online content onto your HDTV. But you might not want to go to all that trouble — at least not at first. Instead, you might decide to tippy-toe your way into the world of online content on the HDTV by just trying out a few downloads and seeing how they work for you.

A simple way to get your feet wet is to simply leverage the DVD burner on your home PC and the DVD player that's installed next to your HDTV. (Just about everybody we know with an HDTV has a DVD player, so we assume you do too!) This approach is sort of like the *sneaker-net* alternative to Ethernet networks that many folks employ in their home PC environment. Instead of setting up a full-fledged network, they use removable media to physically carry content from device to device. You don't even *have* to be wearing sneakers to do so, but you might get the job done a bit faster if you are!

The easiest way to do this is to simply download content on your HDTV and burn a DVD by using programs like iDVD (on the Mac) or Sonic RecordNow! Plus on the PC. Download the content you want to watch, burn it to a DVD-R or DVD+R, and carry it over to the DVD player to watch away.

Keep two things in mind if you want to do this:

- Some content has a copy-protection system in place that keeps you from burning it to a DVD. This situation will occur on a case-by-case basis; check out the fine print on the Web site you're downloading from.
- You need to make sure that your DVD player can actually decode and play the file format you're downloading. Many new DVD players, for example, support the DivX or XviD codec systems (discussed elsewhere in the chapter), which are quite common for Internet content, particularly stuff found on peer-to-peer file-sharing networks. Others support Windows Media and QuickTime formats; many support MPEG-2 files. MPEG-2, after all, is what the DVD itself uses for encoding video files.

A million permutations are out there, so you have to do a bit of research and trial-and-error testing to see what works for you. But because blank burnable DVDs cost about 60 cents a piece these days, the learning experience won't be overly expensive. We do recommend that you buy name-brand blank DVDs from companies like Verbatim or TDK, rather than the supercheapo media you'll find under a cloud of dust in the back of the store. We've had better luck with the name-brand stuff.

The best place we know of for finding specific answers and "how to" information about burning video DVDs is online at `www.videohelp.com`. Keep in mind that this site is a bit complex at times, so we also highly recommend *CD & DVD Recording For Dummies,* 2nd Edition, by Mark L. Chambers (Wiley).

A few DVD players (like the Oppo Digital DV-970HD, `www.oppodigital.com`) also have slots on the front of the player for USB flash memory cards, such as SD cards and Memory Sticks. This lets you skip the burning step and just bring the files over on a flash memory device.

Finally, keep in mind that you can also use this sneaker-net approach to deal with things like pictures — almost all DVD players support JPG picture playback from burned DVDs and CDs.

Try these popular digital media adapters:

- **D-Link's MediaLounge DSM-520:** D-Link has one of the first HD-capable media adapters with this $250 product. It connects to your home network via either Ethernet or 802.11g Wi-Fi network connections and includes an HDMI connection (see Chapter 3 for more on this) to your HDTV. The DSM-520 supports Windows Media and MPEG-4 (using the XviD codec) video playback, and it can output signals up to 1080i resolution over that HDMI connection. Find more information on `www.dlink.com`.
- **Buffalo's LinkTheater PC-P3LWG/DVD:** This product (`www.buffalotech.com`) is very similar to the DSM-520 because it has support for resolutions up to 1080i and support for Windows Media and the DivX codec. LinkTheater throws in a progressive-scan DVD player for good measure. Like the DSM-520, the LinkTheater can connect to your network via both Ethernet and 802.11g. It's about $275 at online retailers.
- **NETGEAR's EVA700 Digital Entertainer:** NETGEAR's new entry in the digital media adapter market supports resolutions up to 1080p using its component-video outputs. With a list price of $330 and street prices considerably lower, the Digital Entertainer can connect to your network by using Ethernet or optional 802.11g or powerline network adapters. The coolest thing about the EVA700 is the fact that it can work hand-in-hand with NETGEAR's SC101 Storage Central *NAS* (network attached storage) server. So you don't even have to keep your PC up and running with the EVA700 — it can pull downloaded and stored movies and videos right off the SC101 without a PC. Find it at `www.netgear.com`.

802.11g has a *theoretical* maximum speed of 54 Mbps. By contrast, an HDTV signal should top out at 19.5 Mbps or so (often considerably less when a more efficient codec like MPEG-4 is being used). So no problem, right? Wrong. The real world speeds of Wi-Fi networks are often a quarter or less of the theoretical maximum. So in all but the best circumstances, you'll have troubles with high-definition content traveling across a wireless network. It might work, and it might not. And even if it does, you might experience drop-outs (where part of the picture doesn't update), blockiness, and even frozen pictures on-screen. Your best bet (until the newer and faster 802.11n system hits the streets some time in 2007 or even 2008) is to use a wired Ethernet connection if you possibly can.

Part III
Movie Machines

In this part . . .

Have you noticed everything in HDTV boils down to acronyms? Well, in this part, we keep up that tradition, talking about DVDs, DVRs, and VCRs, and we do so ASAP PDQ!

DVDs are responsible for the boom in digital video content. The low cost, ubiquitous availability, and sheer massive storage space of DVDs revolutionized everything from blockbusters to direct-to-DVD movies — it has even spawned businesses that rent DVDs via mail, such as NetFlix. So we look at DVDs and DVD players to bring you up to speed on all this great content you can use to drive your HDTV experience. We also talk about the future — great new technology that will bring true HDTV programs to the DVD format.

We discuss the latest and greatest thing to hit consumer electronics: the DVR (digital video recorder). TiVo started the DVR craze, but now you can get DVR functionality in your satellite, cable, and telephone company set-top boxes, too. DVR functionality has transformed how people interact with their TV sets. The DVR lets users watch what they want, when they want, even if they missed it. Got that? You will.

Finally, the VCR is still alive and kicking. If you're like us, you have shelves of VHS movies and older VCRs spread across the house. With new HD camcorders and HDTV content over the airwaves, you might want to upgrade to a new digital VHS VCR. We tell you the pros and cons of D-VHS options, and what's available on the market.

Chapter 11

DVDs

In This Chapter

- Playing today's DVDs
- Making your own
- Combining DVDs with other gear
- Up-converting your DVDs
- Getting into the DVD's high-definition future, now!

DVDs are probably your main source of nonbroadcast video content for HDTV viewing. DVDs have gone from zero to 1,000 miles per hour in no time. More than 70 percent of American households adopted the technology in its first seven years — making DVD the most rapidly adopted consumer technology ever. And now that you can buy a perfectly decent DVD player for under $50, there's no reason to stay on the fence.

The biggest questions about DVD don't have anything to do with today's DVDs and DVD players. You can buy a serviceable DVD player for $30 — if you don't mind getting trampled at Discount City — or a high-quality unit for $150. The confusion is about the *next* generations of DVDs: the profusion of recordable DVD formats and the competing high-definition DVD formats.

A few optical disc players (they're technically not DVD players) on the market today provide true high-definition pictures (and we talk about them in the final sections of this chapter). These new players are great, but pricey, so many folks might want to wait a while before they get one. In the meantime, there's still a lot going on in the world of DVD players, and in particular some standard DVD players are *better* for HDTV. We tell you all about that in this chapter.

In this chapter, we start with the basics of standard DVDs and DVD players and then guide you through the confusing territory of DVD recorders. We follow with all the *form factors* (packages) that contain DVD players — DVD players are in all sorts of electronics gear. We finish by giving you the scoop on the two new high-definition DVD-like formats that are hitting the market as we write — HD DVD and Blu-ray. We even put on our Crystal Ball Spectacles to tell you how we think these high-definition DVD formats will evolve over time.

Discovering the DVD

Digital Video Discs (sometimes they're called Digital *Versatile* Discs, but this usage has pretty much disappeared) are simply *optical* (laser-readable) storage media. Like a CD, the DVD is 12 cm in diameter and consists of *layers.* The top layers protect the DVD, and the inner layers contain tiny pits that a laser beam detects and reads.

Today's DVDs can hold at least 4.7GB of data on one side (compared to about 800MB on a CD). That's enough for about two hours of *standard-definition* (NTSC) video.

Currently, you can put more video on a DVD in two ways:

- **Add a layer:** DVDs can be designed with a second layer of data — basically another layer of *pits* at a different depth within the DVD. The laser in a DVD player can focus on this layer and ignore the other layer entirely.
- **Use both sides:** DVD data can be on both sides of the disc.

  DVDs seldom use both sides of a disc to show different parts of a movie. You usually see double-sided DVDs when

  - A 16:9 *widescreen* version of a movie is on one side.
  - A 4:3 *pan-and-scan* version is on the other side.

DVDs can be both double-layered and double-sided — holding up to 18GB of data.

You can fit more data on a DVD in a couple other ways. We cover both of these at the end of the chapter:

- A new *blue* laser that can use smaller pits
- New data-compression systems

Dealing with Today's DVDs

Today, DVDs usually are the best source for playing prerecorded movies on an HDTV. Here's why:

- **Distortion-free digital image:** It isn't HDTV, but DVDs offer a clear, colorful, and sharp picture at higher resolution than either VHS or most DVRs currently offer (with the exception of D-VHS VCRs, which are almost impossible to find, and a growing body of HD DVRs).

- **Progressive scan:** With a progressive-scan DVD player, your progressive-scan HDTV can be fed a noninterlaced picture (often providing better quality than if your HDTV had to de-interlace the picture itself).
- **True widescreen:** With discs labeled *anamorphic* or *formatted for widescreen,* you can get a full widescreen 16:9 picture for your HDTV without techniques that reduce the resolution of the video, such as letterboxing (some movies are formatted as originally designed with a wider aspect ratio than 16:9, and those still show on the 16:9 screen using the letterbox method — just less of a letterbox than they'd have on a 4:3 TV).

You can find a wide range of prices for good DVD players — from well under $100 up to the thousands for high-end models.

Essential features

We insist on all of the following features in a DVD player for any HDTV system. You should be able to find an affordable (read that as $75 or under) DVD player with

- **Progressive scan** (described in Chapter 21)
- **3:2 pulldown correction** (described in Chapter 21)

  If your monitor has a good 3:2-pulldown-correction system, you can get by without it in the DVD player, but you aren't likely to find a good DVD player without built-in 3:2 pulldown.
- **Component-video outputs** (described in Chapter 3)

Useful features

Depending on how you use your DVD player, these features can make your HDTV system more useful and convenient:

- **Built-in surround-sound decoder:** If your old surround-sound system doesn't have a built-in decoder for current formats, this decoder can add Dolby Digital and DTS (which we cover in Chapter 18).
- **Support for multiple formats or regions:** Many DVD players can support video discs recorded not only in NTSC but also in other standard definition TV formats such as PAL. This is incredibly handy if you're a fan of, for example, European movies that have been recorded in this format. Support for multiple regions is a bit trickier to find, but it is available. Some DVD players come from the vendor with multiple-region

capabilities, or you can hack them by going into the service menu of the player to play discs from any region in the world. Again, this is great if you're a fan of international film and TV or if you live in a country where movies always end up coming out on DVD six months later than the rest of the world. (It happens, and it stinks!) If you're not familiar with regions and DVDs, check out the sidebar titled "Being regional."

- **Extra audio formats:** All DVD players also play *audio CDs,* but some go beyond the call of duty with these audio formats:
  - SACD (Super Audio CD — a high-resolution audio format)
  - DVD-Audio
  - MP3 audio files on homemade CDs
- **Multidisc changer:** A multidisc changer is handy if you watch lots of movies (or use it as a CD player for a party).

  Single-disc players are often more reliable than multidisc changers.
- **HDMI output:** If your HDTV has an otherwise unused HDMI input, you may consider this feature. Some DVD players with DVI-D outputs feature an internal *scaler* (described in Chapter 16) to better match the DVD output to the HDTV. We talk about these players in the section titled "Getting into Up-Converting DVD Players." HDMI is described in Chapter 3.

Being regional

One of the most vexing features that we as consumers have had stuffed down our throats in recent years is the whole concept of DVD *regions.* Essentially, these are arbitrary controls (or one could say, limits) placed upon DVD players by geography. The deal is this: Movie studios often do phased rollouts of movies around the world — where you might have in-theater releases of a movie in one set of countries at the same time as another set is getting their DVD release. The movie content owners don't want to shoot themselves in the foot (in terms of profits) by allowing DVDs imported from that latter group of countries to reduce the in-theater box office receipts in the former group.

Their solution was to impose some electronic restrictions on DVD playback, including two elements:

- A region code on prerecorded DVDs themselves, which designates the geography in which the DVD can be played.
- A region code embedded inside the DVD player. This code must match the region code on the DVD in order to allow playback.

Nine region codes are in use:

- Region Code 0: Allows playback anywhere (This isn't a "real" region code, more of a de facto one.)

- Region Code 1: For the U.S., Canada, Bermuda, and U.S. Territories
- Region Code 2: Western and Central Europe, Baltic States, the Middle East, and a few other regions
- Region Code 3: Hong Kong, Macau, Taiwan, and other parts of Southeast Asia
- Region Code 4: Mexico, the Caribbean, and Central and South America
- Region Code 5: The former Soviet Union, parts of Africa not covered in other regions and the Indian subcontinent
- Region Code 6: Mainland China
- Region Code 7: Reserved for future use (We guess this will cover the lost island of Atlantis, when it's discovered.)
- Region Code 8: A special code for ships, planes, and other international venues

Typically, when you buy a DVD player, it's pre-set at the factory for one (or maybe two) of these regions. A few players are available as Region 0 players right out of the box, but it's most common for users to find instructions on the Internet to enable this feature on regular DVD players by using remote control commands that access the DVD player's hidden service menu. We don't tell you how to do this (because we don't want any copyright enforcers chasing us down), but you can find instructions on the Internet pretty easily if you look in the online HDTV and home-theater forums and communities.

Getting into Up-Converting DVD Players

The real hot spot in the standard DVD market these days is in players designed specifically for displaying standard-definition DVDs on your HDTV screen. These players, typically known as *up-scaling* or *up-converting* DVD players, utilize an internal de-interlacing system (like progressive-scan DVD players) and an internal *image scaler,* which takes the 480 line, interlaced image on a DVD and converts it to a progressive-scan image in a different resolution (720p or 1080i typically, 1080p on a few high-end models) that is the same as (or at least closer to) the native display resolution of your HDTV display.

If all this interlaced and progressive and resolution talk is confusing, skip over to Chapter 21, where we discuss it in greater detail.

Up-converting DVD players have (or should have) two things in common:

- **Scaler circuitry within the DVD player:** This circuitry handles the de-interlacing and the resolution scaling (from 480i to something else) within the DVD player.
- **Digital connections to the HDTV:** Most models use HDMI, though a few models use DVI-D or even RGB (a typical computer connection found on a small number of DVD players — often custom modified models).

Although you might find a few up-converting DVD players that can send the up-converted signal over their analog component-video cables, we typically don't recommend this option. What you're doing in this case is taking the digital video on the DVD, de-interlacing and scaling it, and then converting it *back to analog* to send to the TV, where it is most likely reconverted to digital to run through the HDTV's own internal circuitry and then finally displayed on your screen. All this conversion and reconversion tends to subtract from your picture quality. The only reason we'd even consider using component video for an up-converted signal is if our HDTV simply had no DVI or HDMI ports on it.

Beyond these two basic and common features, you can find a lot of differences between up-converting DVD players. The two main areas of difference are

- **The supported *output* resolutions:** Most up-converting DVD players send a *non*-converted 480p signal over an HDMI connection, or a 720p or 1080i up-converted signal. A few of the most recently announced models output a 1080p signal over an HDMI connection. The important thing to keep in mind about these players is that many "1080p" HDTVs don't accept 1080p as an *input* resolution, meaning that you won't actually gain any benefit from this feature on the DVD player.
- **The *quality* of the up-conversion circuitry:** This is hard to quantify without getting into a player-by-player discussion. The key thing to remember here is that there's a huge range of up-converting DVD players, ranging from $59 models sold at Wal-Mart to $2,000 top-of-the-line models sold by high-end HDTV retailers. Although they all put out a good picture, some of the less expensive models don't use circuitry and systems that are as good at scaling as the better models. With these less expensive models, you might find that the de-interlacing and scaling systems built into your HDTV are actually better at the job than the ones in the DVD player.

As you shop for an up-converting DVD player, we recommend that you spend a fair amount of time looking at reviews and reading the online forums like AVS Forum (`www.avsforum.com`) to see exactly how well the scalers inside the models you are considering work. You might also want to do some trial-and-error testing when you buy your up-scaling DVD player, checking out the different output resolutions and seeing which one matches up best with your HDTV itself. Figure 11-1 shows the popular (and inexpensive) DV-970HD from Oppo Digital (`www.oppodigital.com`). This model includes up-scaling and also SACD and DVD-Audio playback for only $150.

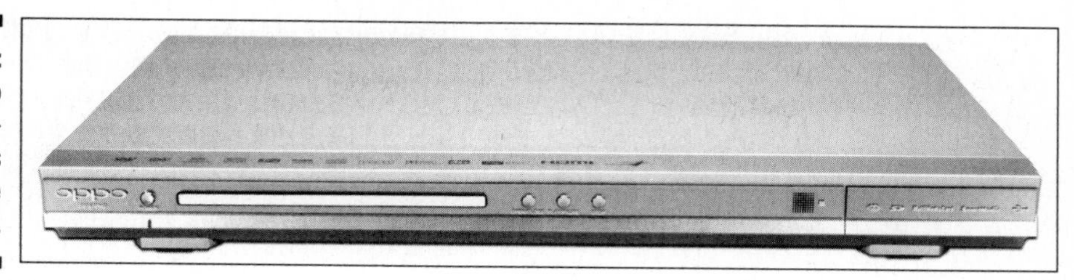

Figure 11-1: The Oppo Digital DV-970HD is inexpensive but capable.

Some folks who use an *external* scaler (discussed in Chapter 16) or who have an HDTV with a really good internal scaling system (for example, many HDTVs use the well-regarded Faroudja DCDi system) purchase an inexpensive up-scaling DVD player that can output non-scaled 480i on the HDMI port. In this situation, you can maintain a digital connection from the DVD player to the TV but at the same time bypass the lower-grade scaling circuits in the DVD player and use the better circuitry in your TV or external scaler.

All the HD DVD and Blu-ray players that we've seen on the market to date support up-conversion for regular DVDs. So there's no need to buy both an up-converting DVD player and a Blu-ray or HD DVD player. Now if only someone would make a universal Blu-ray/HD DVD player (and we've heard many rumors of such a device, but nothing firm as we write), we'd be in one-box heaven!

You might see up-converting DVD players marketed as "high definition-capable" or "ready for your HDTV" or something like that. Remember that although these DVD players can indeed make your picture look better while you're viewing regular DVDs on your HDTV, they are *not high-definition video sources.* The true high-definition optical discs are the Blu-ray and HD DVD systems we discuss in the final sections of this chapter.

Deciphering DVD Recorders

If you could travel through time back to around 1986, you could really blow the minds of anyone you met by saying that less than 20 years later, you can make your own CDs at home. It's like saying you can make your own jet air-liner in the garage. But you can't even buy a PC these days without a CD burner built in.

In less than seven years on the market, the DVD has also become something you can make yourself:

- Most personal computers come with DVD burners.

  Extra capacity makes the DVD format really useful in general with PCs. As quickly as computer hard drives have grown, an 800MB CD now seems puny.

- You can buy a standalone DVD recorder. It's an A/V component that can also be your DVD player.

Recordable DVDs labeled with an *R* can be *written* (recorded) once. RW discs can be *rewritten* (changed) thousands of times.

A DVD recorder can't copy most commercial DVDs, which are copy protected. Your DVD recorder can detect signals in the video stream and prevent copying. You can record TV programs, most VHS tapes, and homemade content (such as camcorder tapes).

The biggest decision to make when considering a DVD recorder is the *format*. The consumer electronics industry (and PC industry) doesn't agree on a standard format for recordable DVDs. You can find three competing DVD recording formats (which aren't always compatible with each other):

- **DVD-R/RW** (DVD dash) discs are the most compatible with other DVD players, so a DVD-R or DVD-RW is most likely to play on your mom's old DVD player.

- **DVD+R/RW** (DVD plus) discs are essentially as compatible as DVDs. (Some DVD+ vendors claim *more* compatibility — we think it's about a draw.)

  DVD+ can record dual-layer discs with double capacity.

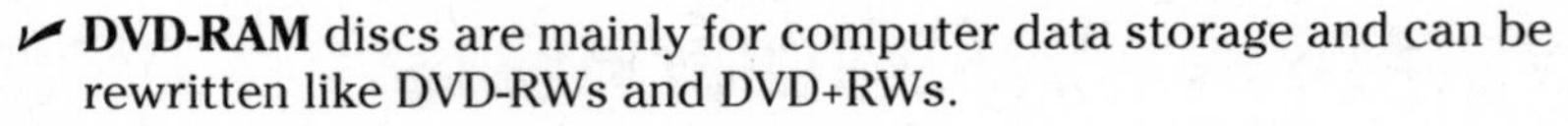

- **DVD-RAM** discs are mainly for computer data storage and can be rewritten like DVD-RWs and DVD+RWs.

  DVD-RAM is the least compatible format with other players.

Finding DVDs in Unusual Places

The most common way to buy a DVD player is to get your hands on a standalone DVD player, but that isn't the only way to get DVD into your home. DVD players have been slotted into all sorts of different home entertainment gear — usually in an effort to save space and money by creating all-in-one devices that can feed different signals into an HDTV.

Xbox/Xbox 360 and PlayStation 2/PlayStation 3 game consoles can be DVD players, too!

Home Theater in a Box

The most common all-in-one DVD sources are the *Home Theater in a Box* (or HTIB) systems from just about every major A/V gear manufacturer. These systems usually include

- An *A/V receiver* (also called a home-theater receiver; see Chapter 18) with a built-in DVD player
- A complete set of surround-sound speakers (including a subwoofer) for Dolby Digital sound from DVDs and HDTV

Figure 11-2 shows an HTIB system.

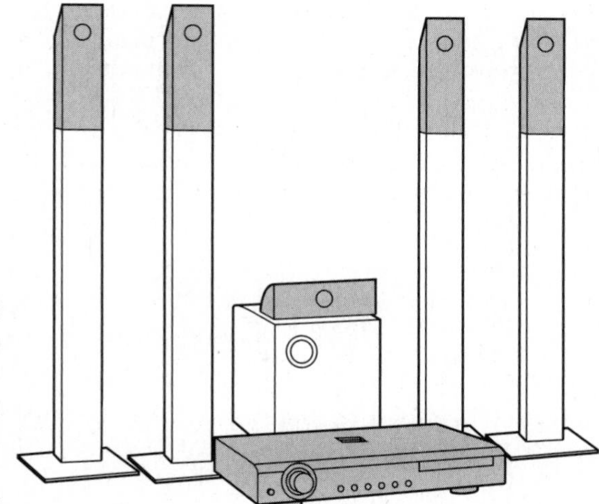

Figure 11-2: Get it all in one box with a HTIB system.

Chapter 18 covers sound systems for HDTV and home theaters.

DVD/VCR

Some manufacturers squeeze a DVD player and a VHS VCR deck into a single chassis.

These devices use one remote control and one power supply, but usually two sets of video outputs that connect to your HDTV:

- Component video for DVDs (plus S-video and composite)

  Avoid DVD/VCR units that don't have component video for the DVD player. You really want be able to use component video to connect the DVD player to your HDTV!
- Composite video for VHS (or S-video if the VCR portion supports that system)

You can't record commercial DVDs onto the VCR in these units, because of a copy-protection system in DVDs.

DVD/PVR

The neatest development we've seen recently is the integration of DVD players (and particularly DVD *recorders*) and PVR (personal video recorder; Chapter 12 covers these devices). It's the latest and greatest for recording TV shows and watching movies — the two replacements for the VCR in one chassis!

With a DVD recorder and a PVR in the same box, it's really easy to make archived recordings of TV shows that you've recorded on the PVR. You don't have to worry about permanently deleting programs if you can just move them to a recordable DVD.

Peering into the High-Def Future

Today's DVDs can't store or play HDTV movies and video, because they were designed when standard definition was good enough. (Actually, it is possible to record HDTV on a regular DVD, but you must use special compression systems, and today's regular DVD players can't handle this, although it might work on a PC.) Three changes are required for HDTV from an optical disc:

- The discs need more storage space than the 5GB of a conventional DVD.

  HDTV programming can have up to eight times as much data as the same show in standard definition.
- The players must read these new high-capacity optical discs.
- The players must output HDTV signals.

Fortunately for HDTV addicts, the major players in the video industry have worked furiously to develop high-definition systems. Unfortunately, these companies have taken separate (and incompatible) paths to DVD HDTV nirvana.

We use the term *optical disc* instead of DVD here on purpose, because these are technically *not* DVDs. They are, however, very similar to DVDs, and you can think of them as high-definition versions of the familiar old DVD.

The path being chosen to increase DVD capacity (or storage space) is to switch to *blue-laser* technology. Blue lasers can read smaller pits in the DVD disc than red lasers (because a blue laser has a shorter *wavelength* than a red laser). Smaller pits mean more pits (and data) fit on a disc. Both of the disc formats we discuss here, HD DVD and Blu-ray, use blue lasers, though they do it in different — and incompatible — ways.

Another way that these new high-definition discs are increasing the capacity to store high-definition video is to use improved compression technologies to store the digital video on the disc. DVDs use a system called MPEG-2, whereas the newer disc formats use newer compression systems like MPEG-4 that can fit more video in the same amount of disc space without a noticeable decrease in quality.

With these competing systems, we expect another *VHS versus Betamax* battle in the market. One format probably will end up "winning." If you pick the wrong system, you'll have obsolete equipment much sooner than you expected. Don't be the first on your block to buy one of these machines (unless you're loaded and don't mind losing the investment). Give the market some time to sort out. We hope that eventually universal players will play both types of discs.

You need a *new disc player* for the following HDTV formats. However, these new standards promise their new disc players also will be compatible with your old DVDs.

Two major competing systems on the market now can provide high definition on an optical disc: HD DVD and Blu-ray. We talk about HD DVD first, simply because it has been on the market longer than Blu-ray.

When you're evaluating these systems, keep in mind these two things: First, the capabilities of the players themselves (and their prices!); and second, the availability of content in each of the formats. This second issue is our main concern. We're sure that the disc player manufacturers will come up with inexpensive and highly functional players over time because without movies to play, a high-definition player is pretty useless. At this point in time, the movie studios are pretty evenly split between the two formats, with some studios supporting only one or the other, and a few supporting both. The industry is still at the very early stages of the high-definition optical disc era, so it's too soon to really say which format will become the most prominent.

Okay, okay. We know you probably want us to make a firm recommendation about whether you should buy a Blu-ray or HD DVD player. Well, the bottom line is we just don't know. Every day we see another analyst or pundit whose opinions we respect come out and declare "Blu-ray is the winner" or "HD DVD is the winner." What we don't see is any consensus from anyone, anywhere — simply a lot of opinions. At this point, we're not spending our money on either one. (But then again, we get to play with review models, so that's easier for us to do!) If you're an early adopter type of person, feel free to choose one or the other now (or both!). But if you're a bit more conservative with your money, we recommend that you wait until one or the other format becomes extremely popular, or until universal players hit the market. Several manufacturers (such as LG Electronics and Samsung) have discussed such models, but none are on the market as we write.

Heading down the HD DVD path

High-Definition DVD (HD DVD) is a blue-laser system supported by Toshiba, NEC, Sanyo, Microsoft, Intel, and the DVD Forum (an industry group that promoted the original DVD format).

The technical details

The HD DVD format uses discs that are very similar to standard DVDs — with the same 12-cm diameter and the same internal construction (known as the *substrate*) but with considerably smaller pits than a standard DVD. Because the pits are smaller than those on a regular DVD, you can fit more of them on the same-sized disc — and therefore more data. In fact, an HD DVD can hold more than three times the data of a standard DVD, with capacities of 15GB for a single layer disc and 30 for a dual-layer disc (compared to 4.7 and 8.5GB for single- and dual-layer DVDs). To read these smaller pits, HD DVD players use a blue-laser system.

As we write, no dual-layer discs are on the market, but they're on the way. Also on the way are three-layer discs holding 45GB of movie data and *dual-sided* discs that can be flipped over to hold as much as 90GB of data. Even though these higher-capacity discs aren't available yet, the players shipping today will support them when the discs hit the market.

HD DVDs have larger pits in the disc than those in the Blu-ray discs we discuss in the next section. These larger pits should make HD DVD discs easier to manufacture (they can be made on existing DVD production lines with only minor changes), but they hold less data than Blu-ray discs. To make up for it, the HD DVD folks use more aggressive *video-compression* technologies (like Microsoft Windows Media) so longer movies fit on a disc. Compression basically shrinks the amount of storage space needed by a video by discarding "unnecessary" bits of the video. This can cause some degradation in picture quality, but how much you lose is subjective, and it depends on which compression system is used. A good compression system is nearly unnoticeable.

Some other notable features in the HD DVD format include

- **Support for surround-sound format:** These formats include Dolby Digital and DTS along with the newer (and better-sounding) Dolby Digital Plus, Dolby TrueHD, and DTS-HD, which we discuss in Chapter 18.

  Not all HD DVDs include these new surround-sound formats, but the players support them if they're included.

- 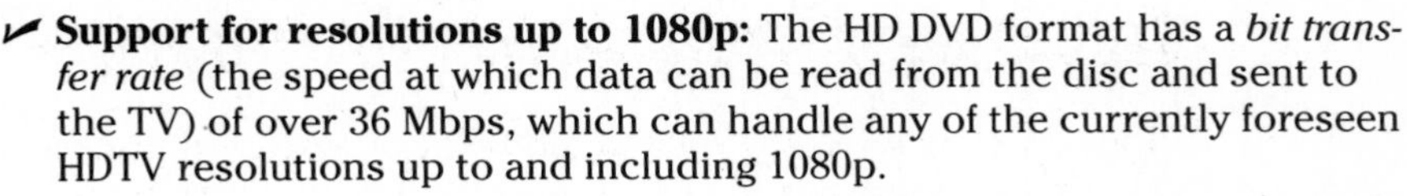

  Support for resolutions up to 1080p: The HD DVD format has a *bit transfer rate* (the speed at which data can be read from the disc and sent to the TV) of over 36 Mbps, which can handle any of the currently foreseen HDTV resolutions up to and including 1080p.

  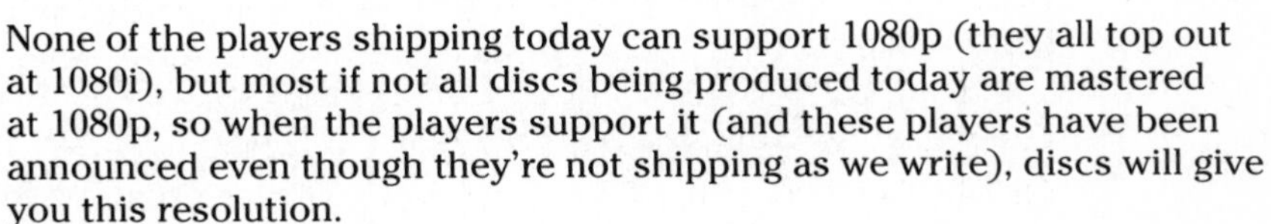

  None of the players shipping today can support 1080p (they all top out at 1080i), but most if not all discs being produced today are mastered at 1080p, so when the players support it (and these players have been announced even though they're not shipping as we write), discs will give you this resolution.

- **Support for advanced video-compression systems:** These systems, like MPEG-4 AVC and Microsoft's VC-1, enable HD DVD to cram a lot of movie onto a single disc without any noticeable decrease in quality. HD DVD also supports the older MPEG-2 compression system used in standard-definition DVDs (and for most broadcast HDTV systems).
- **Support for recordable discs:** These include both one-time recordable (HD DVD-R) and rewriteable discs that can be recorded upon multiple times (HD DVD rewriteable). So far these capabilities are built into PCs only, but we expect standalone recorders to hit the market eventually.
- **Future interactive online capabilities:** The HD DVD platform is designed to allow HD DVD players with Internet connections to access online interactive content related to the content on the disc itself. Though not widely available yet, this system (called iHD) is part of the HD DVD standard and is a feature available on all HD DVD players. All that's needed now is for the content creators (like the movie studios) to implement this interactive programming.

The players

As we write, Toshiba is the only manufacturer shipping HD DVD players. Toshiba offers two models:

- **The HD-A1 HD DVD Player ($499.99):** The first mainstream high-definition player to hit the market, Toshiba's HD-A1 is actually built around a very powerful (Pentium 4) PC architecture, using a customized version of the Linux operating system. The HD-A1 supports both 720p and 1080i outputs (though not 1080p) using HDMI connections to your HDTV. The HD-A1 also has a built-in decoder for Dolby Digital and DTS, as well as the new Dolby Digital Plus, Dolby HD, and DTS-HD Lossless surround-sound formats (for discs including these formats). Like all HD DVD players, the HD-A1 can also play regular DVDs (with an up-converting feature) and regular audio CDs. Figure 11-3 shows the HD-A1.
- **The HD-XA1 HD DVD Player ($799.99):** This player has all the features of the HD-A1 model, but it includes some additional bells and whistles such as a motion-sensitive remote control, an Ethernet port for access to Internet content, and even cool stuff like a motorized door that hides all the working parts on the front of the player but then opens itself when it's time to change discs.

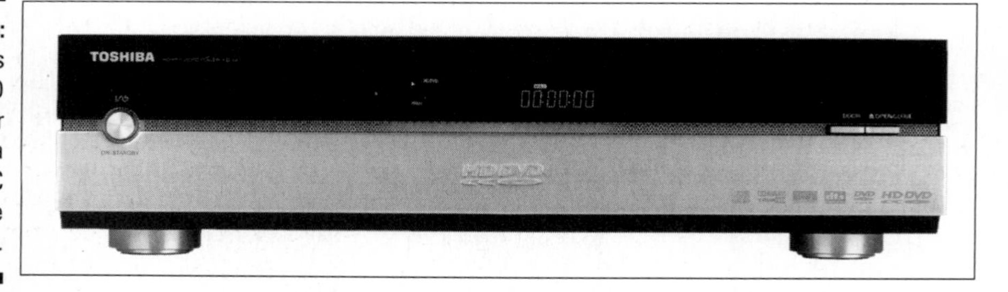

Figure 11-3: Toshiba's HD-A1 HD DVD player is actually a powerful PC under the hood.

As this book goes to press, Toshiba has announced (but not yet started shipping) updated versions of the HD-A1 and HD-XA1 — the HD-A2 and HD-XA2. These models will add 1080p outputs over their HDMI connections. The price will remain the same for the entry-level HD-A2, and will bump up to $999.99 for the fancier HD-XA2.

RCA has also announced an HD DVD player (priced at around $500), and other vendors such as LG and Samsung will soon release players as well.

The interesting dark horse in the HD DVD race is Microsoft. The company as a whole has announced support for HD DVD (meaning you can expect built-in support for HD DVD playback and recording in Windows Vista, due out in 2007). Microsoft has also announced an HD DVD drive accessory for the popular Xbox 360 gaming console (discussed in Chapter 14). Although the HD

DVD drive isn't available on the market as we write, we do know that this will be an inexpensive (around $200) accessory that will connect to the console via USB and will provide a cheap way of getting into the HD DVD game. With an HD DVD drive attached, an Xbox 360 will be pretty close in function (and in price).

The movies

As we write, not a lot of HD DVD titles are available on the market, but the number is growing day by day. The following movie studios have stated support for HD DVD format:

- Universal Studios (exclusively HD DVD)
- The Weinstein Company (exclusively HD DVD)
- Paramount Pictures
- Warner Brothers
- Studio Canal

You can find an up-to-date list of all HD DVD titles at the following URL:

```
www.hddvdprg.com/
```

Look on this page for the HD DVD Contents List button. (It's a PDF file.)

The popular online DVD rental site, Netflix (www.netflix.com), has HD DVD discs available as part of its service, and Netflix states that it has *all* HD DVD titles as they become available. If you're not familiar with Netflix, we recommend that you check it out: The service provides you with a certain number of discs for a monthly fee. (The number of discs varies depending upon your service plan.) Netflix doesn't charge late fees. You simply choose movies online, and they're sent to your house. When you're finished, you mail them back in the Netflix prepaid mailers. As soon as the movie is back at Netflix, you automatically are mailed the next movie in your Netflix queue. Pretty cool concept, we think!

Being Blu-ray

Competing with HD DVD is a large group of companies supporting a blue-laser system called *Blu-ray.* The Blu-ray bandwagon includes Sony, Panasonic, RCA, Pioneer, Philips, Samsung, LG, and many more. (You can find the full list at the Blu-ray Disc Association's Web site: www.blu-raydisc.com.)

Because it uses smaller pits in the disc than HD DVD, a single-sided Blu-ray disc can hold up to 25GB of data on a single layer (50GB on a dual-layer disc), which is more than enough to handle most movies in HDTV resolution. A dual-sided version of Blu-ray can expand these capabilities up to 100GB.

The Blu-ray disc itself *appears* to be very similar to a DVD or HD DVD (and it comes in the same 12-cm size), but its internal structure is significantly different than that of an HD DVD or regular DVD, which means that it requires a pretty significant investment in special production lines to create the movie discs that you rent or buy — a bit of a disadvantage for Blu-ray because the added expense makes movie studios more likely to go with HD DVD.

Some features of the Blu-ray system include

- **Support for the highest HDTV resolutions, up to and including 1080p:** Like the HD DVD system, Blu-ray supports all standard- and high-definition resolutions up to 1080p. Unlike HD DVD, the first Blu-ray players hitting the market already support 1080p — so if you have a 1080p HDTV that can support 1080p *inputs* on its HDMI connection, this might be a good reason to choose Blu-ray over HD DVD.
- **Support for the advanced surround-sound formats (Dolby Digital Plus, Dolby TrueHD, and DTS-HD) is included in the standard:** Dolby Digital Plus is mandatory (meaning that all players and discs must support it); Dolby TrueHD and DTS-HD are both optional, so support is on a case-by-case basis.
- **Support for the advanced compression systems like MPEG-4 and Microsoft's VC-1:** Note that so far, all commercial Blu-ray discs use the older and less efficient MPEG-2 compression system (taking away some of the advantage of Blu-ray's additional disc space, because video encoded using this system takes up more space per minute).
- **An interactive capability for viewing both disc-based and Internet-based content related to that on the Blu-ray disc itself:** The Blu-ray system is based upon the Java programming language and is called *BD-Java.* Not all Blu-ray disc players have this capability, but many do.
- **Support for regular DVDs with up-scaling:** All Blu-ray disc players (like all HD DVD players) can play regular standard-definition DVDs, and the players up-scale them to 720p, 1080i, or even 1080p.

  Not all Blu-ray disc players play regular audio CDs, but we expect that over time they all will.

- **A special coating on the disc to eliminate skipping and playback errors from scratching:** Because of the especially small pit size, Blu-ray discs might be even more susceptible to damage from scratches and improper cleaning. The Blu-ray folks have come up with a system called *Durabis,* which is a hard polymer coating on the surface of the disc that is supposed to withstand even steel wool! As we write, it's not clear whether all commercial prerecorded Blu-ray discs use this coating. This coating is both a benefit and a potential pitfall for Blu-ray — if the coating is there, you have sturdier discs, but if it's not, you need to be especially careful when handling or cleaning your discs.

The players

A few manufacturers are making Blu-ray disc players as we write:

- **Sony BDP-S1 ($999.99):** Sony is the lead company behind the Blu-ray disc effort, so it's no surprise that one of the first generally available players comes from the company. The BDP-S1 is a 1080p-capable Blu-ray disc player that can also up-convert regular DVDs to 1080p through its HDMI output. (It also supports 720p and 1080i, so if those resolutions are better for your HDTV display, use them.) The only beef we have with the BDP-S1 is its inability to play regular audio CDs. You need a stand-alone CD player in your home-theater system if you want to do that.
- **Samsung BDP-1000 ($999.99):** Samsung is another of the major players in the Blu-ray effort (and a frequent partner of Sony's — they work together in the HDTV display business and have established a gigantic technology cross-licensing deal to help each other develop new products). The BDP-1000 was the very first Blu-ray disc player on the U.S. market, and like Sony's DBP-S1, it supports 1080p playback of Blu-ray discs, as well as up-conversion to 1080p for regular DVDs, but you must use the HDMI connection to get 1080p playback of either type of disc. The BDP-1000 *does* support the playback of audio CDs as well, so it can take the place of an audio CD player in your home-theater set-up. Figure 11-4 shows the BDP-1000.

Figure 11-4: Samsung's BDP-1000 is the first Blu-ray disc player in the U.S.

Other models due out soon include Panasonic's DMP-BD10 ($1,299.95) and Pioneer's BDP-HD1 ($1,500), and soon to follow, many more models from dozens of manufacturers. But the really big event in the Blu-ray world will come when the Sony PlayStation 3 (or PS3, as everyone calls it) hits the streets in late 2006. Every PS3 will include a built-in Blu-ray disc drive, and the Premium models will have the capability to play Blu-ray video discs in full 1080p high definition over an HDMI connection the HDTV. In fact, Blu-ray will be the format for all of the PS3 game data (there's a data-only variant of Blu-ray, just as there is of HD DVD and regular DVD), which will help support the economies of scale required to get Blu-ray production lines up and running around the world.

We talk about the PS3 in more detail in Chapter 14. The thing to remember here is that a $500–$600 investment in a PS3 gaming console will also get your HDTV into the Blu-ray video world!

The movies

Roughly the same number of movie studios support Blu-ray as those that support HD DVD — meaning that there's not a huge advantage for one over the other so far. The biggest difference is that more studios support Blu-ray *exclusively* than those that support HD DVD exclusively. Some studios have hedged their bets by supporting both systems, whereas others have taken a strong stand on one side or the other (and most of those have come down on the Blu-ray side of things).

Among the studios supporting Blu-ray for their movies are the following:

- 20th Century Fox (exclusively Blu-ray)
- Metro-Goldwyn-Mayer (exclusively Blu-ray)
- Sony Picture Studios (exclusively Blu-ray)

  Both MGM and Sony Pictures are owned by Sony, so it's no surprise that they support Blu-ray exclusively.
- Paramount
- Warner Brothers
- Lions Gate Home Entertainment
- Walt Disney

Because HD DVD players arrived on the market first, currently (in the summer of 2006) more HD DVD titles are available than are Blu-ray titles, but over time this might shift. You can find a listing of available Blu-ray titles by going to the Blu-ray Disc Association's Web site (`www.blu-raydisc.com`) and navigating to High Definition Movie Releases in the Products section of the Web site.

Chapter 12

Getting Into DVRs

In This Chapter

- Throwing away your VCR
- Controlling TV your way
- Going high-def
- Looking into DVR options

Besides HDTV itself, we think that the rise of the *DVR* (digital video recorder — some folks call them PVRs, or personal video recorders, too) is the single biggest thing to happen to TV in our lifetimes. Yeah, we know that sounds like an exaggeration, but trust us — using a DVR is almost a life-changing experience. After you start using a DVR, you'll never (never ever) think about TV the same way again!

In this chapter, we give you a good dose of DVR background and information, and then we cover the (few) DVRs that can record HDTV signals. We also discuss DVRs more generally — after all, most features and goodies found on today's standard-definition DVRs will eventually appear again in high-def versions of the same units. Finally, we cover some interesting DVR variants (such as the DVR/DVD recorder combo that Danny finds so compelling).

In this chapter, we focus mainly on *stand-alone* DVRs (that's to say, DVRs that are individual components designed to record TV). Many DVRs these days are being built into other things like cable set-top boxes, satellite receivers, and even PCs. We mention those as well, but get into details about those types of DVRs in Chapters 8, 9, and 10, respectively.

Some of the DVRs in this chapter (and most DVRs you can buy today — at least the ones you can buy off the shelf at the local electronics store) *can't* record or play HDTV content. A very small number of HDTV-capable DVRs are available, but we expect that very soon HDTV DVRs will be common.

DVR 101

A DVR is, at the most basic level, a digital replacement for your VCR. Just like a VCR, a DVR can record TV programs. Instead of using a tape for this purpose, a DVR records TV digitally, on a computer hard drive inside the DVR.

Discovering the benefits

If the only benefit of a DVR was the replacement of the tape with a hard drive, we'd be sold on it. But as they say in the Ginsu knife commercials "Wait, there's more!" DVRs can also

- **Pause live TV:** Usually, DVRs automatically record about an hour's worth of the show you're currently watching. So you can hit the pause button, run to the bathroom or kitchen or answer the phone, and pick up later where you left off.
- **Rewind live TV:** You can rewind a scene that you missed. If you simply have to watch Jerry Springer's most recent move on *Dancing With the Stars* again, a DVR lets you do so.
- **Record a show you're watching with one touch:** If you must leave the room suddenly, just press a single button on the remote to save the whole show (including the part you've already watched).
- **Record an entire series:** You don't have to set up separate recordings for each episode of a series. Just select one episode and turn on series recording. Your DVR automatically finds every episode of that show on the programming guide and records it for you.

All DVRs have a few common elements:

- The hard drive for storing video
- A *GUI* (graphical user interface) that lets you control the DVR with your remote control
- An on-screen programming guide for scheduling recordings (no more need to manually enter dates and times!)
- A telephone or Internet connection for updating the programming guide

The best-known DVRs come from a company called TiVo (`www.tivo.com`) and its partners. In fact, the name TiVo is widely used to refer to DVRs in general. It's even used as a verb as in, "I didn't watch it yet, but I TiVo'd *Deadwood.*" By the way, the folks at TiVo prefer that you don't genericize their brand name — which, unfortunately for them, many people do. We avoid that here by using the term DVR, except when we're talking about a TiVo-brand DVR.

Making the connection

Connecting a DVR to your HDTV system is simple. You want your DVR to be *inline with* (connected between) your TV source (antenna, cable, or satellite) and the HDTV itself. Here are some suggestions on hooking up a DVR:

- This is incredibly easy if the DVR is part of the TV source (either a satellite receiver or cable set-top box). There's no hookup to do — just connect.
- If you have a *stand-alone* DVR, you connect it to the standard-definition outputs of your satellite or cable set-top box (or to your antenna cable) before those signals reach your HDTV.

Today, only one HDTV stand-alone DVR is available on the market, so chances are you have only standard-definition signals on your DVR. That's because the common, self-contained DVRs that are most popular today haven't yet been converted to handle HDTV — though this is changing rapidly. Although we've not seen them in person yet, the release of TiVo's Series3 DVRs in late 2006 will be the most prominent example of HD-capable stand-alone DVRs because these stand-alone models will accept both over-the-air and cable HDTV signals.

You could use one of the DBS satellite HD DVRs we discuss in the following section to record over-the-air HDTV signals. In this case, the DVR is also your HDTV tuner — the HD DVR satellite receivers available from DISH Network and DIRECTV include ATSC over-the-air TV tuners for picking up your local HD channels.

Figure 12-1 shows a typical DVR setup with a satellite or cable set-top box and a VCR.

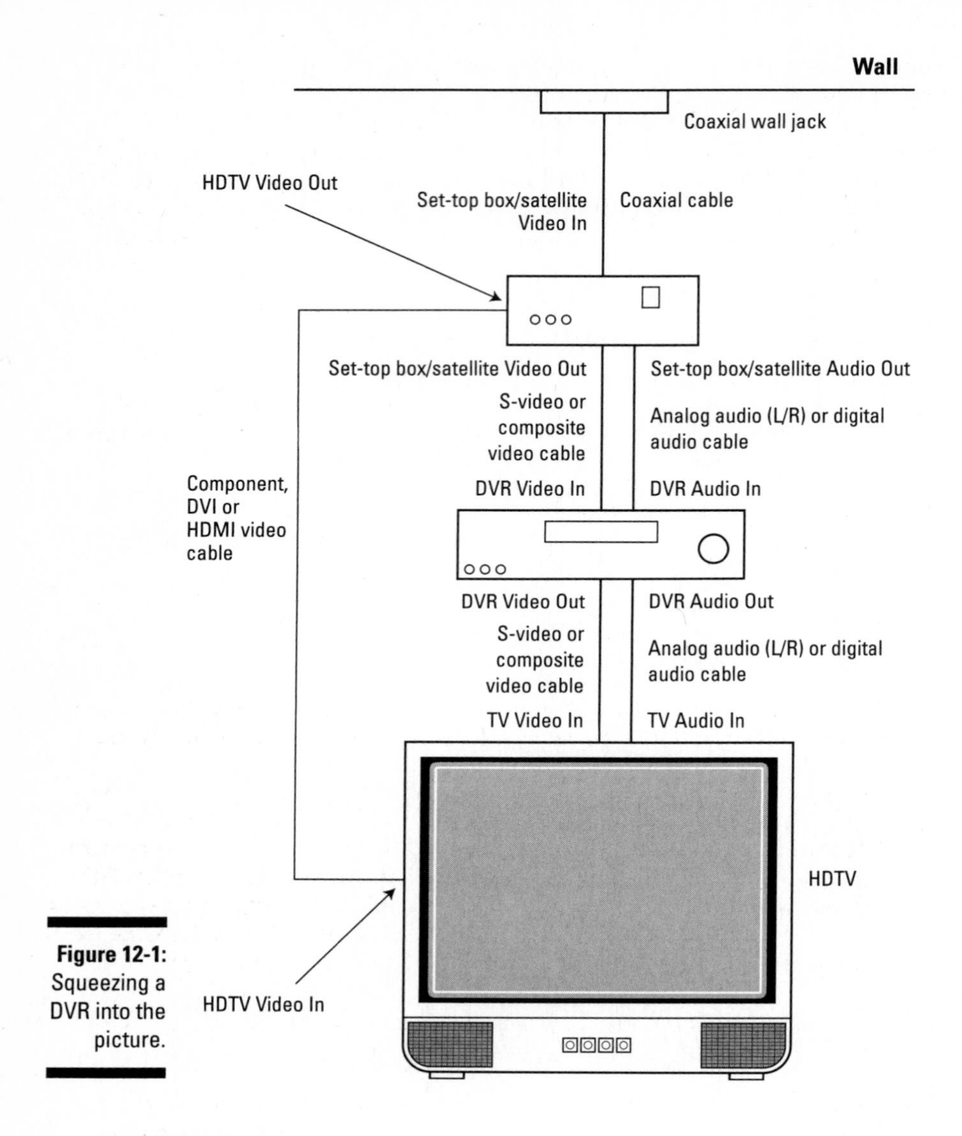

Figure 12-1: Squeezing a DVR into the picture.

HD DVRs

If you own or plan on owning an HDTV (and we assume you do, because you're reading *HDTV For Dummies*), your ultimate DVR records and displays HDTV programs in their native HDTV resolutions (720p or 1080i). After all, your favorite HDTV shows deserve the deluxe DVR treatment as much as your standard-definition favorites.

As we write, few HDTV-capable DVRs are on the market. Making a high-def DVR is a cost-engineering challenge, not a technical challenge. (The technology is available, but it's not cheap!) Today, a high-def DVR isn't easy to build at prices that consumers will pay, because a high-definition DVR requires

- A much bigger hard drive
- A more powerful graphics chip
- More expensive connectors (like DVI-D or component video)

The prices of high-def DVR components are coming down rapidly. We expect that HDTV DVRs will soon be common and affordable.

For cable and satellite

If you want an HDTV DVR today, you have a couple choices:

- **From your cable company:** A few cable companies offer digital cable/HDTV set-top boxes that include HDTV DVR features. One example is Scientific Atlanta's Explorer 8300HD, which you can look into at the following site:

  `www.scientificatlanta.com/consumers_new/CableBoxes/8300hd.htm`

  You can't buy one of these for yourself, but if you're lucky, you can rent one from your cable company for a monthly fee. The Explorer 8300HD (shown in Figure 12-2) contains a 160GB hard drive that can record up to 20 hours of HDTV programming (or 90 hours of standard-definition programming, or between 20 and 90 hours of the two combined).

  Not all cable-company "HDTV" DVRs actually record in HDTV — some combine an HDTV tuner with a *standard* DVR. This DVR is better than nothing, but know what you're getting for your money.

- **In your satellite receiver:** Each of the major satellite TV providers have announced deals to provide customers with set-top boxes that can record HDTV. We talk about the satellite service providers (DIRECTV and Dish Network) in Chapter 9, and we describe your options for HD DVRs there.

Figure 12-2: The Scientific Atlanta Explorer 8300HD can record HDTV.

Moving up to Series3

The exciting news that many HDTV enthusiasts have been waiting for is due to become a reality in late 2006 as TiVo launches its Series3 DVR system. The Series3 DVR will include all of TiVo's well-loved features (like Season Pass) and integrate them into a new hardware system that can handle both over-the-air and digital cable HDTV signals (using the CableCARD system we discuss in Chapter 8).

The Series3 TiVo is a powerful high-end design and is priced accordingly, at $799. Inside the box, you can find the following features:

- **Two ATSC tuners:** These tuners let you pick up over-the-air HDTV broadcasts (see Chapter 7) from two separate local broadcasters. These tuners also pick up NTSC (analog) broadcasts, so basically, if your antenna can pick it up, the Series3 can tune it in, display it on your HDTV, or record it.
- **Two CableCARD slots and QAM tuners:** Designed for connecting to a digital cable service, the QAM tuners let you pick up standard- and high-definition signals being provided by your cable service provider. The CableCARD slots allow you to rent one of these cards from your cable company to *decrypt* (or unscramble) premium services (like HBO or other pay channels). See Chapter 8 for more info on QAM tuners and CableCARD.

Your cable company must support the CableCARD system for this to work. Most do, but unfortunately it's still kind of hit-and-miss, so verify this before you shell out the money for a Series3.

- **An HDMI output for connecting to your HDTV:** The Series3 TiVo includes an HDMI output for connecting to your likewise-equipped HDTV (or even to an HDMI-outfitted home-theater receiver, if you have one). This gives you an opportunity to maintain an all-digital signal all the way to the TV, for the highest picture quality. The Series3 also includes analog component-video outputs in case your HDTV doesn't have any available HDMI slots.

If your HDTV has DVI inputs but not HDMI, you can use an inexpensive HDMI-to-DVI adapter, which can even be built right into your HDMI cable to make the connection. Remember that your TV needs to support HDCP (the copy-protection system) for this HDMI-to-DVI conversion to work.

- **Coaxial and Toslink digital audio outputs:** You can connect these to your home-theater receiver for decoding the Dolby Digital 5.1 audio signals included on many HDTV broadcasts to feed your home theater's surround-sound system.
- **Lots of hard drive space:** The Series3 TiVo will include room for recording up to 25 hours of HDTV programs, 300 hours of standard definition programs, or some combination of the two. (That is, if you have a mixture of HDTV and SDTV recorded, you'll be able to save somewhere between 25 and 300 hours, depending upon how much of each type you have recorded.)
- **Memory expansion:** The Series3 includes a *SATA* (Serial ATA) expansion port designed to accept connections from SATA external hard drives. With such a hard drive (these are different from the USB and FireWire hard drives that you might be familiar with), you can easily expand the storage capabilities of your DVR to record and archive additional programming.
- **Built-in Ethernet and available wireless:** You can connect a TiVo to your home network and broadband connection. With this connection, you can connect to TiVo's online portal, as well as other Internet services like Yahoo! Photos, and also for connection to other TiVo boxes in your home using TiVo's home networking features.
- **TiVoToGo:** TiVo's TiVoToGo feature lets you transfer shows you've recorded from the TiVo itself to PCs and other devices for viewing anywhere around the house or on the road. With TiVoToGo and TiVo's desktop (Windows XP or 2000) or mobile device software (available for Sony PlayStation Portable, Palm Treo, Apple iPod, Nokia N80, Toshiba Gigabeat, and Creative Zen Vision M), you can watch your shows on your desktop or laptop PC or on any of these portable devices.

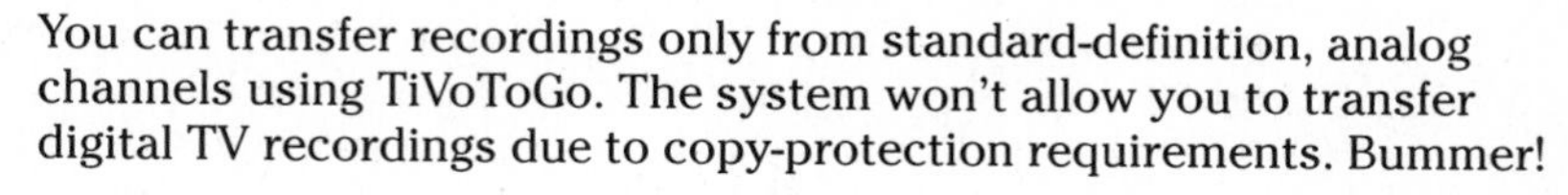

You can transfer recordings only from standard-definition, analog channels using TiVoToGo. The system won't allow you to transfer digital TV recordings due to copy-protection requirements. Bummer!

This is all topped off by TiVo's famous interface and user-friendly features like Season Pass (which lets you set up recordings of an entire season's worth of a show with one quick push of a button). Figure 12-3 shows the Series3 TiVo.

Figure 12-3: Get your HDTV recorded with a Series3!

Standard DVRs

If you're one of the lucky few without a budget, the right HDTV service provider, and the right alignment of the stars, you can get an HDTV-capable DVR *right now.* For the rest of us, a standard-definition DVR is the only option until HDTV DVRs become more widely available (and affordable!).

What to look for

DVRs are made by many manufacturers and are available in a wide range of prices. Here are the key things to look for when deciding which DVR is right for you (these criteria can apply to an HDTV DVR, too):

- **Purchase or rent:** Your main question probably is, "How much is this thing going to cost me?" The answer depends upon where the DVR comes from. Some units are offered to consumers for outright purchase; others are rented from a cable company or other provider on a monthly basis.

  Do the math on this — rental is initially cheaper but might be more expensive if you keep the DVR for a long time.

- **Programming guide fees:** Many DVRs require a monthly service fee to access the on-screen program guide. (It makes the DVR work, and you should consider it mandatory.) Some, like those rented from a cable company, might not charge for this or might include this cost in the equipment rental fee. In some cases, you can save money by purchasing a "lifetime" subscription.

  Lifetime subscriptions are a good deal only if the company that sells them stays in business for the lifetime of your DVR. DVR manufacturers have gone bankrupt. Folks with lifetime subscriptions weren't affected, but for a while it wasn't clear what would happen. If your DVR vendor goes bust, you and your subscription could be out in the cold.

- **Hard drive size:** The main technical specification to check in a DVR is hard drive size — often described as *hours of programming* that fit on the hard drive. Hours can trick you — they depend on video *quality.* Compare the same video quality when comparing time.

  Check whether the DVR allows for adding external hard drives to add capacity in the future. Look for a DVR that includes either *FireWire* (also called 1394) or *USB 2.0* connectors for this purpose. Read the manual's fine print — some DVRs have these connectors, but they aren't "turned on," so you can't use them for this purpose.

- **Number of tuners:** We recommend a two-tuner DVR, so you can either record two programs at once or watch one while recording another.

  Some DVRs have only a single TV tuner built in. You can record only one program at a time. If this single tuner DVR is built into your cable or satellite set-top box, you might be able to watch only the program you're recording.

- **Hardware features:** The most basic DVRs simply have connections for TVs and TV source signals. Others can connect to home networks (via Ethernet) or even wireless LANs for sharing content between DVRs within the home. We recommend Ethernet connections so you can share! You can also use this Ethernet and broadband connection for downloading program guide information, instead of tying up a phone line.

- **Software features:** Some DVRs (like those made by TiVo) are built around very sophisticated GUIs that can do things like recommend shows you'd like (based on your viewing habits), display pictures, play music that's stored on your PCs, or even incorporate Internet content. Others (like some of the DVRs incorporated into cable set-top boxes) are more basic.

  If you want all the fancy interface stuff, you're probably best off with a TiVo-powered DVR.

Finding a DVR that fits

When buying a DVR, you need to consider how it fits into the space you want to put it in — its *form factor.* Some DVRs are self-contained units — in other words, a chassis with a DVR inside, and nothing else. Others are incorporated into other devices, providing a multipurpose device that takes up less room on that equipment rack under your HDTV.

Among your choices are the following:

- **Stand-alone DVRs:** The most popular DVR is a stand-alone unit, like the traditional TiVo and ReplayTV models.

  These DVRs are best suited for connecting to over-the-air broadcast TV or analog cable systems, though you can use them with a digital cable set-top box or satellite receiver.

- **Cable and satellite DVRs:** If you connect your HDTV primarily to digital cable or a satellite service, you should consider a DVR that is built into your receiver or set-top box. These DVRs save space and receive the *digital* signals used by these services (which stand-alone DVRs can't do). Remember that satellite users usually *buy* the DVR/receiver and cable users usually *rent* equipment from the cable company.
- **DVD recorder DVRs:** Our favorite stand-alone DVRs have a built-in DVD recorder. A DVD recorder solves the biggest issue most DVR owners run into — what to do with all those back episodes of your favorite show when your hard drive starts to get full. With a DVD recorder, you burn a DVD archive and clear your hard drive for new recordings.
- **Using your PC:** There are a couple of ways to use a PC:
  - If you have a Windows XP Media Center PC, you *have* a DVR. These PCs already have the hardware and software to record TV programs; the stuff works just like a regular DVR.
  - You can turn a standard PC into a DVR by adding some inexpensive hardware and software.

  Chapter 16 covers the equipment you need if you want to use a PC as a DVR.

Using your DVR on the 'Net

TiVo, one of the originators of the whole DVR concept, has been facing an uncertain future as DVRs are built into more and more devices (such as cable services' set-top boxes and Media Center PCs). Indeed, the company faces a bit of a squeeze as cable and satellite companies develop their own DVRs that might be more convenient for cable and satellite customers than using a stand-alone TiVo.

One way TiVo is fighting back is by turning to the Internet. TiVo has announced that it will offer a new service to its *broadband* (cable- or DSL-modem-equipped) customers. An Internet service will allow customers to download movies, video, and music directly to their TiVo DVRs. Basically, this service is designed to both supplement and to *bypass* regular TV sources — providing TiVo with a compelling service of its own to keep people buying TiVo DVRs (and to keep them subscribing to TiVo services). Today, you can already program your TiVo online and access services such as Yahoo! Photos; soon you'll be able to download videos and shows directly from the Internet for playback on your HDTV. Keep an eye out for this service's availability at `www.tivo.com`.

Chapter 13

Taping Time

In This Chapter

- Viewing your old *Scooby Doo* VHS tapes on your HDTV
- Making the VCR-to-HDTV connection
- Understanding the advantages of digital VHS
- Buying your first (and only) D-VHS VCR

In our collective rush to sift through the $5 DVD bins at Wal-Mart, it's easy to overlook the good ol' VCR that sits precariously atop many of our TV sets at home. DVDs are cooler, packed with more data, and they're . . . well . . . digital.

But almost everyone has a lot of investment in the VCR — ranging from the scores of VHS files purchased over the years to the recorded broadcast and home movies that make up your VHS collection. It might not be a big investment dollarwise, but you might have an emotional investment in those old tapes! Even in the face of several generations of higher-quality video options — laser disc, DVDs, satellite, and now HDTV and DVRs — it's hard to part with the dependable VHS VCR.

That's because the venerable VCR is simply a useful device — and it has even been remaking itself for this new, high-definition world. In this chapter, we talk about how to get your old VCR hooked up to your HDTV so you can dust off those old taped episodes of *LA Law* (you '80s fan, you!). We also discuss some of the (few) new digital VCR models remaining on the market that you might want to investigate to upgrade your taping capabilities.

Checking Out Your Digital Options

If you stroll into your neighborhood electronics store or surf for VCRs online, you can find three kinds of VCRs available today: VHS, S-VHS, and D-VHS.

Familiar to all, VHS VCRs are the standard VCRs you've seen in stores for almost 30 years. These VCRs use VHS videocassettes and record a low-resolution TV signal — about 240 lines of resolution (a third of the 720 lines you get with HDTV!). These VHS VCRs can come cheap (less than $50) or expensive (more than $800); most include stereo audio capabilities (if so, they're labeled *HiFi*) and analog Dolby surround-sound capabilities.

You might not be so familiar with S-VHS or D-VHS VCRs, so we discuss them in depth in the following sections.

S-VHS VCRs

The S-VHS (or Super VHS) VCR provides a higher-quality, higher-resolution image — 400 lines (a lot compared to the 240 you get with VHS VCRs). In the midst of the craze for more resolution, you might guess that film buffs jumped at the first S-VHS VCRs when they were introduced, but they've actually been slow to take off, for several reasons:

- With the early models, you had to buy more expensive (and harder-to-find), specialized tapes to record in S-VHS mode.
- You can't play an S-VHS tape in a regular VCR (but you can do the reverse).
- S-VHS VCRs themselves have been significantly pricier than the standard-format VCRs on the market.
- Very few prerecorded movies were available in the S-VHS format.

Lately, however, S-VHS VCRs have made a bit of a comeback, driven by lower equipment pricing — S-VHS VCRs start off at around $120 — and a new technology boost in the form of S-VHS-ET (Expansion Technology), introduced by JVC. S-VHS-ET allows an S-VHS VCR to record a full S-VHS signal on regular VHS tape. Higher-resolution source material from camcorders, satellite, and digital cable transmission — and the ensuing display of recorded tapes on higher-definition displays — have raised the bar for video recorders as well. (You really notice poor resolution when it's 50 inches and 1,280 x 720 pixels in front of you!)

There's very little difference in pricing between a high-quality VHS VCR and an S-VHS VCR. (Really cheap VHS VCRs are a lot cheaper, of course.) If you're already spending 100 bucks or so for a HiFi VCR, we think you should find the extra $20 to get into an S-VHS model.

You also find a few VCRs on the market that are labeled *Quasi-SVHS.* These units *can* play S-VHS tapes, but only at the lower VHS resolution. They're handy if you don't care about the higher resolution but have some S-VHS (or S-VHS-C camcorder) tapes that you want to watch.

D-VHS VCRs

If you're going whole-hog high-def, you want to look at a high-definition VCR — specifically, the D-VHS VCR. D-VHS VCRs can play and record standard VHS and S-VHS formats. More importantly, D-VHS VCRs can also record and play all the common HDTV formats (discussed in Chapter 1), including 1080i, 720p, 480p, and 480i — though not in 1080p, which is not widely available but is beginning to come onto the market with the advent of the HD DVD and Blu-ray disc players.

To record and play HDTV, the D-VHS system needs a FireWire (also called i.LINK or IEEE 1394) connector to connect to an HDTV TV or a stand-alone HDTV tuner. Many HDTVs and HDTV tuners don't have a FireWire connection, however, because broadcasters and movie studios don't want you to be able to record HDTV programs. So be sure of your connection options before plunking down the cash for a D-VHS VCR.

The D-VHS VCR sounds like a great idea; indeed, if you tape a lot, this might be the VCR for you. However, you're probably going to be disappointed with its use on a day-to-day basis, for a couple of reasons:

- D-VHS would be pretty cool, if many people weren't restricted from using it for its intended purpose — recording HDTV shows for later playback. Unfortunately, there's a fundamental disconnect between the HDTV connection provided on D-VHS decks (FireWire) and that found on the majority of HDTV tuners/set-top boxes (DVI).
- Only a limited number of movies is available on prerecorded D-VHS videocassettes — truly a small number (less than a hundred) compared to what you can get on DVD (tens of thousands).

Many people who used to use VCRs a lot are focusing on DVRs (see Chapter 12) or are using DVDs, particularly as HDTV-capable DVD formats are moving from the announcement stage and are finally available on the market (see Chapter 11). As a result, D-VHS VCRs have not really taken the market by storm — and really dropped in price recently as a result. (They can be had for about $600.) The D-VHS VCR category is largely heading for Laserdisc status (that is, dead), or at least confined to a niche market of D-VHS enthusiasts.

What will really nail the coffin of the D-VHS system is the advent of HDTV-capable DVD systems — particularly when these offer recording capability to users. Combine a high-def DVR with a recordable Blu-ray DVD system, and you have the best of all worlds. No tapes to mess with, just pure, digital, skip-right-to-where-you-want-to-be-without-rewinding bliss.

Given that this is a book on HDTV and not on VCRs (and that the main high-definition angle to VCRs is D-VHS), we spend the rest of this chapter talking about how to hook up your VCR into your HDTV system for best viewing. We also look at one of the only D-VHS products actually on the market as we write. If you want to know more about VCRs in general, try `http://hometheater.about.com`.

Connecting VCRs to HDTVs

You often have a choice of connections/cable types to use when connecting source devices (such as VCRs) to your HDTV (Chapter 3 covers the details). Choosing which of these connections to use can be (initially) a bit confusing, but they fall into a hierarchy, and when you get a handle on that hierarchy, matters get easier: Just pick the best connection option available on both ends of the connection.

When it comes to VCR connection options, we can make it even easier for you. Your choices depend on what kind of VCR you own:

- **Connecting VHS VCRs:** If you have a VHS VCR (not an S-VHS model), you have only two choices: You can connect by using a composite video cable (and a pair of audio cables), or you can use the coaxial cable output of the VCR. We recommend that you use the composite video cable — it delivers a much better picture than the coaxial cable.
- **Connecting S-VHS VCRs:** These VCRs add an *S-video* connector (That's where the name S-video came from originally — even though S-video connections are now found on DVD players, set-top boxes, game consoles, and other source devices.) S-video provides a much better picture than composite video, so use this connection (and a pair of audio cables) to connect your S-VHS videocassette recorder.

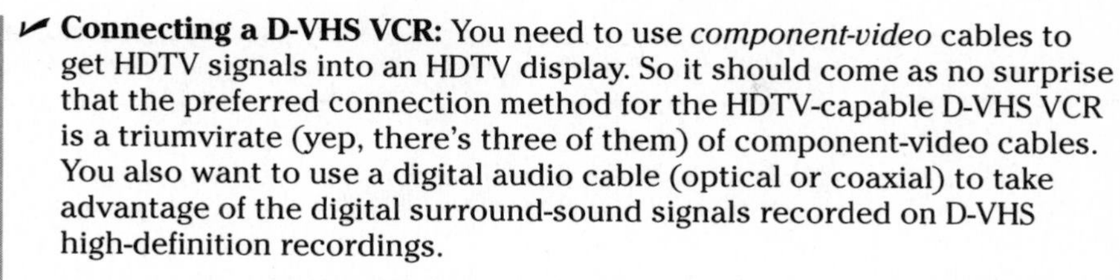

- **Connecting a D-VHS VCR:** You need to use *component-video* cables to get HDTV signals into an HDTV display. So it should come as no surprise that the preferred connection method for the HDTV-capable D-VHS VCR is a triumvirate (yep, there's three of them) of component-video cables. You also want to use a digital audio cable (optical or coaxial) to take advantage of the digital surround-sound signals recorded on D-VHS high-definition recordings.

  If you don't use the component-video cable, you can't view HDTV recordings in high definition.

In addition to these playback connections that enable you to view videocassettes played on your VCR, you also want to make a recording connection between your TV source and your VCR. Your options are many and depend greatly on a whole host of variables regarding what else you've connected to your HDTV (and how it's connected). Here are some general rules to follow:

- If you have a D-VHS deck and you want to record in HDTV, the answer is simple: You need to connect your HDTV tuner to the D-VHS by using the FireWire (or i.LINK) cable. You have no other option — which severely limits the usefulness of the D-VHS because few HDTV tuners have functioning FireWire connections that make this work.
- For a non-HDTV VCR recording, your options depend on two things: whether you've connected a home-theater receiver as part of your overall HDTV home-theater system, and whether you require a set-top box or satellite receiver for non-DTV TV channels.
  - If you're using a home-theater receiver (discussed in Chapter 18), take advantage of this device's *video-switching* capabilities. You can connect all your video source devices (such as DVD players, cable set-top boxes, and satellite receivers) into a set of *inputs* on the receiver. Then you can use the *outputs* on the back of the receiver to connect both your HDTV (for viewing) and your VCR (for recording). This is the best and most flexible way of connecting your VCR for recording.
  - If you don't have a home-theater receiver, you have to connect your VCR *inline* between your TV source and your HDTV. This could be as simple as plugging an antenna cable into the back of your VCR (to the *TV In* connection) and then using another antenna cable (coaxial cable) to connect the *TV Out* jack on your VCR to the antenna input on your VCR. If you have a set-top box or satellite receiver, you need an extra set of S-video or composite-video cables to make this connection. Figure 13-1 shows a VCR installed inline.

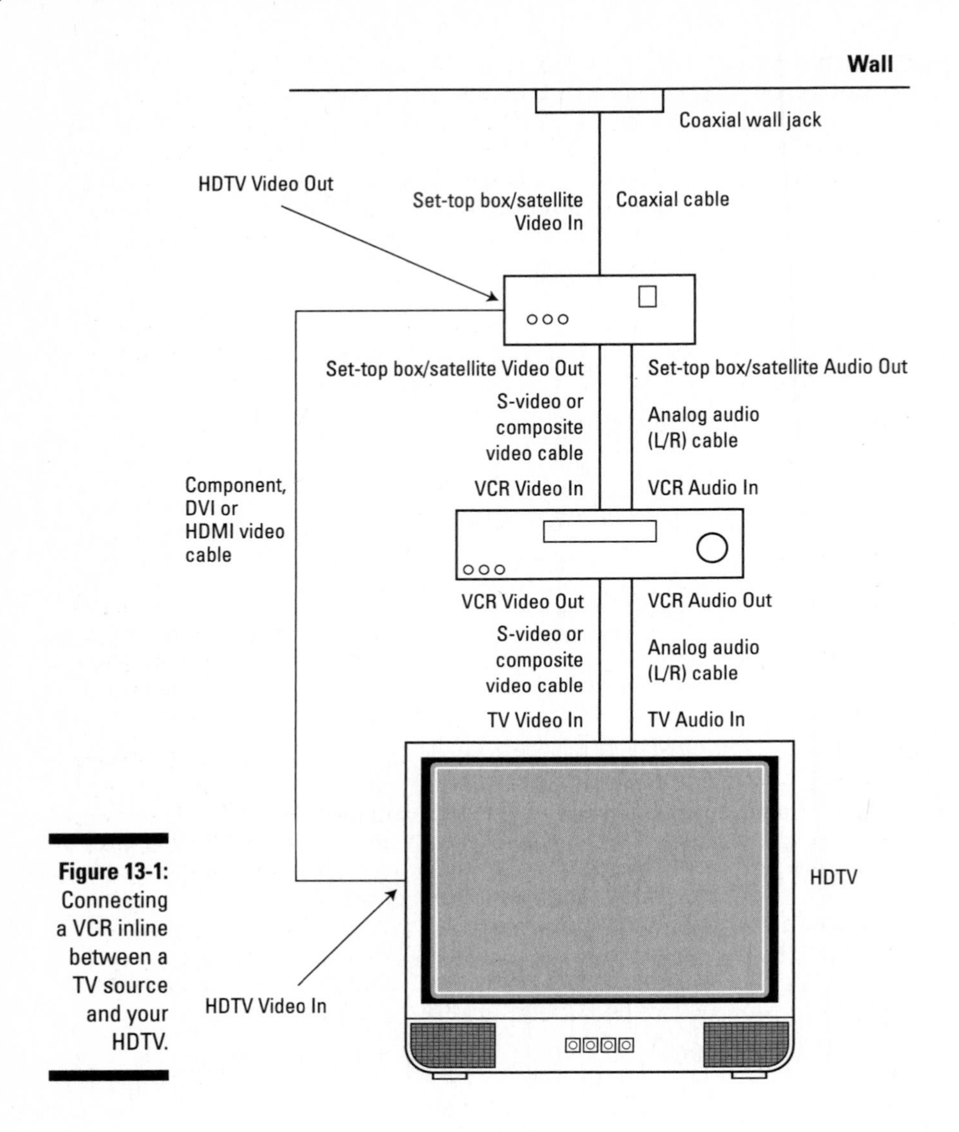

Figure 13-1: Connecting a VCR inline between a TV source and your HDTV.

Looking Closer at D-VHS

We discuss D-VHS in a bit more detail because it's an interesting format and because it's currently the only system that lets you record HDTV programming to archive. (DVRs let you record to the hard drive, but you can't get the HDTV recordings off the DVR to save them, or at least you won't be able to until HD DVD or Blu-ray recorders hit the market.) But — and this is an important *but* here — D-VHS has pretty much been deemed a market failure, and manufacturers are no longer developing new products, and in many cases manufacturers have dropped them entirely. So if you want to get into D-VHS, your best bet is to look for an overstock/discontinued/used recorder on eBay or craigslist.com. Mitsubishi still shows its D-VHS units on its Web site (`www.mistubish-tv.com`), as does JVC (`www.jvc.com`) but none of the other manufacturers do.

So if you have high-definition leanings (and why else would you be reading *HDTV For Dummies*?), you at least owe it to yourself to consider whether a D-VHS will work for you. So . . .

- **Here's what you can expect to do with it:**
  - Record over-the-air broadcasts
  - Watch D-VHS prerecorded tapes
  - Store HDTV camcorder content to D-VHS
- **Here's what you might not be able to do with it:** Record cable or satellite broadcasts. (Many set-top boxes lack a FireWire port to send digital signals to the D-VHS, due to copyright concerns — and some that have the capability don't have it enabled by the software, so it's a useless connection that doesn't work.)
- **Here's what you can't do with it:** Make tapes of your favorite Hollywood DVD movies. A copy-protection system called Macrovision keeps you from doing so, and it's built into all DVDs and all VCRs. Even if you want to copy your favorite movie from DVD to VHS to play on a VCR in your vacation home or somewhere else, you're just plain out of luck.

If these benefits and limitations seem reasonable to you, consider getting a D-VHS VCR. When it comes to playing and recording high-def signals, you have to go D-VHS or you're simply not getting all you can out of your HDTV lifestyle! No other format can match those high-res images . . .

. . . at least not yet. The new high-def DVDs and players that have just hit the market have resolutions that equal those of D-VHS.

The D-VHS VCRs use high-capacity D-VHS tapes to record HDTV broadcasts at their full resolution. They can up-convert or down-convert HD recordings on playback to match your TV's display rate (even when connected to a regular analog TV).

D-VHS VCRs typically can do some other handy tasks:

- Record/play Digital VHS, Super VHS, Super VHS ET, and standard VHS formats.
- Store (on one DF-420 D-VHS tape) up to 3.5 hours of material in HS mode (28.2 Mbps), 7 hours in STD mode (14.1 Mbps), 21 hours in LS3 mode (4.7 Mbps), or 35 hours in LS5 mode (2.8 Mbps).

  The higher the *bit rate* is (the preceding Mbps numbers), the less compressed the video is, and the better it looks.
- Record/play at 1080i, 720p, 480p, and 480i resolutions.
- Record/play 5.1-channel Dolby Digital and DTS HDTV broadcasts for surround sound.
- Play HiFi stereo VHS and S-VHS for Dolby Pro Logic surround sound.

You can find some other VCR features from regular VHS/S-VHS VCRs that you might recognize. For example, you might find VCR Plus+, which allows an average human being to record a certain show at a certain time more easily. With this system, all you have to do is punch in a special code that is listed next to the show you want to record in your paper's TV listings (or in *TV Guide*). More advanced versions of VCR Plus+ (called Silver and Gold) allow you to localize your VCR Plus+ settings for your area.

VCR Plus+ is handy, but after you've used a slick DVR interface (such as TiVo), you'll never want to go back to using even the finest VCR system! Of course, you can record HDTV on TiVo only if you buy a $1,000 DIRECTV TiVo system.

At the time of this writing, we could find only one or two D-VHS products left on the market (and we looked a lot!). One of the most popular and prominent is JVC's HM-DT1000U (see Figure 13-2). So if you want one, we advise you to move quickly because these might not be around much longer.

Figure 13-2: JVC's HMDT1000U has the D-VHS market just about cornered!

Part IV

HDTV Gadgets Galore

"There's a slight pause here where the PS2 sequences your DNA to determine your preferences."

In this part . . .

Who said HDTVs had to be all about watching DVDs and high-definition satellite signals? We sure didn't. In fact, you have all sorts of other ways to get high-definition signals on and off your HD system — so many that we created this special part to talk about them.

We discuss all the high-definition implications of your Sony PS2/PS3, Xbox and Xbox 360, Nintendo, or other gaming console, and we note which consoles allow you to play DVDs (or even HD-DVD or Blu-ray discs) on your HDTV. Now, *that's* double duty.

We talk about what to do with your camcorder if you want to view your home movies in 1,280-x-720-pixel mode and what features to think about when you crave a new high-definition camcorder for those really fine-looking shots of your kids in pumpkin-head Halloween outfits.

The fun really begins as we discuss the wide range of gadgets that are popping up to accentuate your HDTV purchase. We talk about souped-up PCs designed to support your HDTV-powered home theater, video jukeboxes that make watching another movie truly point-and-click, and how to take a bath with your HDTV. (Shocking isn't it? Oh, bad pun!)

It's gadgets galore in HDTVdom. . . . Is that cool or what? (Now *this* is reason why you really want an HDTV, isn't it?)

Chapter 14

Gaming Consoles

In This Chapter

- Getting into gaming
- Playing with the PlayStation
- Getting high-def with Xbox
- Going nuts with Nintendo
- Making the right connections

Video game consoles and HDTVs can be great together. The *graphics engines* (the chips within the console that create video images) in today's video-gaming machines are incredibly powerful and capable of very high resolutions. The newest video game consoles (the Sony PS3 and the Microsoft Xbox 360) can support full HDTV resolutions. Although not a lot of games on the market offer a resolution that's higher than standard-definition (480 lines), you *can* find many widescreen, progressive-scan games that look awesome on your HDTV screen.

What systems do we have? Pat loves his Xbox, and Danny's kids stick with their PS2 . . . all connected with component-video and Toslink cables into our HDTVs. We're both anxiously awaiting PS3's launch, so we can make our next decision (between PS3 and Xbox 360).

Understanding Consoles

If you're of a certain age (like us), you remember the original Atari consoles and games like Pong — crude black-and-white graphics only. If you haven't kept up with video games since then, you'll be shocked at the amazingly realistic graphics that today's video game console systems produce.

Today's market-leading video game consoles — the Sony PlayStation 2 and 3, the Nintendo GameCube and forthcoming Wii, and the Microsoft Xbox and Xbox 360 — are actually powerful computers that are specially configured to play video games. In fact, these consoles are so computer-like that they can be set up (with official or unofficial software) for such cool nongame functions as surfing the Web, running Linux programs, and checking e-mail.

Some high-end TV aficionados look down their noses at game consoles. Don't let this attitude (which we think is dying off) bug you at all. We think game consoles are perfectly fine for HDTV.

The latest game consoles have a few key features:

- **Powerful graphics engines** can often provide true HDTV (such as 720p and 1080i formats, and even 1080p in the case of the new Sony PlayStation 3), provided the game uses high-resolution video.
- **S-video and component-video connections** bring higher-quality video into your television. PlayStation 3 even includes an HDMI connection (on the more expensive premium model).
- **Optical drives (usually DVD)** store game software, which means more storage space for better graphics. You can also use these drives to play DVD movies. The Sony PlayStation 3 includes a Blu-ray disc drive for playing high-definition movies, and Xbox 360 will soon have an optional accessory HD DVD drive for the same purpose. (See Chapter 11 for more on Blu-ray and HD DVD.)
- **Surround-sound formats** (such as Dolby Digital) provide a truly immersive gaming experience.

If an image stays still for too long on-screen, it can cause *burn-in* — the phosphor coating on the screen stays illuminated with a certain image long enough to become permanent. (Hey, that's why screen-savers were invented.) Susceptible TVs, such as plasma screens, CRT projection systems, and — to a lesser degree — CRT direct-view TVs, can suffer burn-in from excessive gaming. When you have burn-in, you see a persistent (and sometimes permanent) ghost image on-screen of whatever has become burned in. Certain images tend to be in the same place all the time when you play a game (like your race car's speedometer or your ammo indicator), so you could experience burn-in if you put in lots of gaming time. You can pretty much avoid this problem by simply

- Setting up your TV properly (brightness *down*)
- Not gaming for hours and hours every day

LCD (liquid-crystal display) screens, LCD projectors, and DLP TVs are immune to burn-in. They can take as much gaming as you (and your sore thumbs!) can throw at them. Keep the display type in mind if you're still shopping for your HDTV.

Meet the Consoles

You might have an older gaming console (or three) lying around the house — such as an original PlayStation or a Super NES system. By all means, if you'd like, connect these to your HDTV. You might just have one old game that you can't help but pull out and play every once in a while. (For Pat, it's the old Nintendo version of *Dr. Mario* that he just can't give up.) Just keep in mind that these older consoles (typically using an RF modulator or — at best — a composite-video connection) won't rock your world on an HDTV. Aside from the primitive graphics, the resolution is limited to standard-definition specs.

We focus our attentions in this chapter on the two most recent generations (the sixth and seventh generations — gaming consoles have been around for a while) of the major gaming consoles from Sony, Nintendo, and Microsoft because these give you the best results with your HDTV. Each company has two consoles that we discuss: the sixth-generation models (which are on their way off the market but are already in millions and millions of homes) and the seventh-generation consoles (which are either new to the market or not available as we write).

Going with the PlayStations

Sony has manufactured some of the most successful and popular gaming consoles ever. In particular, *PlayStation 2* (or PS2) has sold over 100 million units worldwide and has somewhere in the range of a *bazillion* games available. (Okay, the real number is just somewhere between 5,000 and 6,000, but really, that's a heck of a lot of games.) PS2 is still on the market as we write and probably will be for a while, but it will soon be replaced by an awesomely powerful console called *PlayStation 3* (or PS3).

Playing with PS2

As with all the consoles we're discussing, PS2 (that's what all the cool kids call it) is really a powerful computer, based on a 300 MHz CPU (central processing unit). Just 300 MHz might not sound fast (compared to a 4 GHz Pentium 4 PC), but PS2 has some other stuff that makes it great for games, including these goodies:

- **A powerful graphics chip:** Called the Graphics Synthesizer, this chip — not the CPU — does all the heavy lifting to create the gaming images that show up on your HDTV screen.
- **A DVD-ROM drive:** PS2 games are burned onto DVDs. You can also use this drive to play regular DVDs and CDs, and even older PlayStation (the original version) games.

- **Ports and connections:** PS2 is loaded with ports and connections for attaching peripherals — including connectors for game controllers and USB peripherals, a slot for memory cards, and an i.LINK (FireWire) port for connecting multiple PS2s together for head-to-head gaming.
- **An optional network kit:** For the ultimate in PS2 gaming, you can buy a network kit that provides an Ethernet interface for connecting PS2 to a home network and the Internet. Online gaming is the ultimate goal here — the ability to play with people thousands of miles away!

PS2's graphics capabilities could theoretically support HDTV resolutions, but the system is designed for lower resolutions. With the optional component-video cables — and a game that supports it — you can have progressive-scan gaming (480p) on your HDTV.

PS2 can support Dolby Digital 5.1 channel surround sound (as long as you use a digital audio cable to connect PS2 to your A/V receiver or surround-sound decoder). Most games, however, provide only two-channel stereo sound.

The back of a game's packaging has the Dolby Digital logo if a game supports Dolby Digital.

You can pick up a new PS2 for as little as $129.99 list price.

Stepping up to PS3

PS2 has been around since March of 2000, so it's getting a bit long in the tooth (though there have been some upgrades, mainly to make the console smaller and cheaper). Sony is well aware of this and has been pulling out all of the stops to develop a mightily powerful HDTV-capable gaming console known as PS3.

PS3 is built around the Blu-ray disc format (instead of DVD or some proprietary disc format), which offers a ton of room for game programs and data — and also offers the ability to play high-definition Blu-ray disc movies. The main features and specs of PS3 include

- An entirely new type of CPU called the *Cell processor.* Jointly developed by Sony, IBM, and Toshiba, the Cell processor contains eight individual processing units and is capable of operating many times faster than traditional processors at the same *clock speed.*
- A graphics engine known as the *RSX* (Reality Synthesizer). Developed with the folks at NVIDIA, the RSX operates at 1.8 *trillion* FLOPS (floating point operations per second), and can support full 1080p HDTV resolutions.

- The aforementioned Blu-ray disc drive, which supports both data discs (for gaming purposes) and Blu-ray movies.
- Networking capabilities including Gigabit (1,000 Mbps) Ethernet, Bluetooth, USB 2.0, and optional Wi-Fi for connecting to peripherals and to your home network and broadband Internet connection.
- All the video and audio connections you need to hook up to your HDTV, including composite, S-video, component video, analog, Toslink, and coaxial audio outputs. The premium version of PS3 (more on that in just a moment) also includes a 1080p-capable HDMI output for connecting both audio and video over a single cable.

When PS3 hits the market in late 2006 (it's scheduled for a November 2006 release date), it will come in two different versions:

- **Basic:** This model includes a 20GB (and upgradeable) hard drive for storing gaming info, photos, sounds, and wireless controllers (which take advantage of the Bluetooth connection on PS3), and will retail for $499. Note that this model won't have an HDMI port, so you won't be able to connect your HDTV.
- **Premium:** This $599 model will include a bigger hard drive (60GB, also upgradeable), a memory card reader (for transferring data to the console from devices like digital cameras or music players), Wi-Fi (for connections to a wireless home network), and — most importantly — an HDMI output for connecting to your HDTV.

You can upgrade the Basic version to the Premium version after the fact (by buying accessories from Sony) in all ways but one — you can't retrofit the HDMI connection to the Basic model. For that reason, we recommend that if you're buying a PS3, you dig through your sofa cushions to find the extra $100 if you possibly can.

The Basic version of PS3 *does* support 1080p resolutions through its component-video outputs. So if you can't get the HDMI version, not all is lost. Two "gotchas" to remember here though:

- Not all HDTVs can support 1080p *inputs* on their component video inputs.
- It's *possible* (although not likely in the next couple of years) that gaming software providers and Blu-ray movie creators will use copy-protection systems to lower the resolution available on the component-video outputs (because there's no way of keeping those analog outputs from being copied as there is with HDMI outputs). The technical capability to reduce this resolution is there, but no one yet knows whether it will be used.

X (boxes) mark the spot

Although Sony has been the sales leader in the video console market for the past decade or so, they haven't been without competition. The biggest (and fiercest) competitor that Sony has faced during this decade has been Microsoft, who set up a whole new division and spent billions and billions of dollars developing its own gaming consoles.

Today's reigning king, Xbox 360

The best high-definition gaming console you can buy as we write is Microsoft's Xbox 360. (Recall that PS3 has not yet hit the market, so we haven't seen one in person as we write.) Xbox 360 has been available since November of 2005 and has already sold over five million units worldwide.

Among the many features of the Xbox 360 are the following:

- A powerful *triple core* IBM PowerPC processor operating at a 3.2 GHz clock speed. This processor is more powerful than all but the fastest desktop or laptop PC CPUs.
- A graphics engine from ATI that's over four times as powerful as the one in the original Xbox, and which supports HD resolutions up to 1080i.

  Microsoft *requires* all game developers to create games for Xbox 360 that are both widescreen (16:9) and written for 720p or 1080i HDTV resolutions.
- An optional 20GB hard drive for storing game info, photos, and sound.
- Composite, S-video, and component-video outputs for connecting to your HDTV, and a Toslink (optical) audio cable for connecting to your surround-sound receiver for Dolby Digital surround sound.
- Built-in Ethernet and optional Wi-Fi (a $99 accessory) networking capabilities for connecting to your home network and broadband Internet connection. Microsoft runs an online gaming network (Xbox Live, at `www.xboxlive.com`) that lets you play games against friends and enemies online (or you can use Xbox Live to meet those future friends and enemies).
- Media Center Extender capabilities are built into Xbox 360. What this means is that you can use your Xbox 360 to remotely control and access media (such as recorded TV shows, music, videos, and photos) that are stored on your Windows XP Media Center Edition PC; see Chapter 16 for more on this. All you need is a wired (Ethernet) or wireless (Wi-Fi) connection to that PC.

Coming soon to the Xbox 360 platform is an external HD DVD drive. This drive will connect to one of the Xbox 360's USB ports and will provide you with a means of using the Xbox 360 as a high-definition movie player for your HDTV. Pricing for this HD DVD drive hasn't been set as we write, but rumors of a price around $200 are pretty common.

Xbox 360 has two variants:

- **The Xbox 360 Core System:** This is a cheaper variant, without the hard drive or wireless controllers (it uses wired controllers) and without the Ethernet or component-video cables. It retails for $299.99 in the U.S.
- **The Xbox 360 System:** This is the high-end model, with a retail price in the United States of $399.99. This model includes wireless controllers, an external 20GB hard drive, Ethernet cables for connecting to a home network, and a set of component-video cables for connecting to your HDTV.

You can buy all the extra bits and pieces that make up the Xbox 360 System for your Core System after the fact, if you'd like.

The Xbox 360 doesn't have an option for an HDMI connection to your HDTV, so component video is your only HDTV connection option for high-def gaming.

The first HD game console, Xbox

The previous king of the gaming consoles for HDTV was the Microsoft Xbox. That's because Xbox was the first console that

- Supported true HDTV outputs (720p or 1080i):
  - Just connect the picture via component-video cables, choose the right games, and you're in HDTV gaming heaven.
  - It supports Dolby Digital 5.1 channel surround sound, and many games support this directly.
- Has (at least a handful of) games that support HDTV.

  It isn't just the console that gives you an HDTV gaming experience — the game *software* must support HDTV, too. This support means loads of painstaking work for all the engineers and artists creating the game, so just a few HDTV games are on the market as we write. We expect that this number will grow — particularly as the seventh-generation game consoles reach the market.

Xbox is built around a PC-style architecture. (This probably isn't surprising, given that Microsoft is involved.) The heart of the system is a fast CPU from the Pentium III family. The graphics are handled by a scorchingly fast chip dubbed the X Chip, built by NVIDIA (another big PC graphics chip maker). Other key components of the Xbox include the following:

- **A built-in 8GB hard drive:** Instead of using relatively small memory cards, Xbox has a generous-size hard drive for saving game-state information. You can also rip your own CDs onto this hard drive to create your own game soundtracks.
- **A built-in Ethernet port:** Although networking kits are optional for PS2 and GameCube, every Xbox includes this Ethernet port for connecting to the Internet — and to online gaming via a home network with a broadband cable or DSL connection.
- **A DVD optical drive for games, CDs, and video DVDs:** This drive allows you to play DVDs and CDs with your Xbox. (Keep in mind, however, that the remote control that enables DVD playback is optional.)

The original Xbox is being phased out with the advent of Xbox 360. Although many retailers are still selling Xbox consoles, this console's days are numbered — so stock is limited to what's at hand. If you want an Xbox and don't already have one, you might need to pick up a used one on eBay or another auction site. Microsoft has sold over 20 million Xbox consoles, so they're not exactly in short supply!

Gaming with Nintendo

Nintendo is one of the pioneers of the gaming console — if you're like us, you probably have some fond memories of your Super NES console and hours of fun with Super Mario. Nintendo is still in the gaming console market, but it hasn't been as successful as Sony and Microsoft in recent years. Nintendo's game consoles are smaller, cheaper, and simpler than their Sony and Microsoft counterparts and have been very popular for younger children.

Getting cute with GameCube

The current Nintendo game console is GameCube. GameCube is slightly less powerful than PS2 or Xbox, but it makes up for this by being the cheapest console on the market at $99.99 retail. GameCube is also amazingly compact (6 inches or less in all directions).

Nintendo (in what we assume is a cost-saving measure) has, as of the summer of 2004, removed component-video connection capability from new GameCubes. Many games support 480p progressive scan, but you won't get that through a new GameCube. If progressive scan is important to you, consider looking for a *used* GameCube on eBay or elsewhere. Look for models made before May of 2004 — and *don't* purchase GameCubes with model number DOL-101.

So what's inside a GameCube? How about all the following good stuff crammed into that small package?

- **A powerful graphics processor:** Jointly designed by ATI (big in the PC graphics chip market) and Nintendo, this chip (codename: *Flipper*) can crank out up to 12 million polygons per second — which is a lot of video on-screen.
- **40MB of system memory:** For storing gaming code.
- **Four ports for gaming controllers:** So three friends can come over and play head to head.
- **Two memory card slots:** For storing game state information, so you can start back up where you left off.
- **A proprietary optical disc that's smaller than traditional DVDs or CDs:** In other words, no double duty as a DVD player for GameCube.

Many GameCube games can provide non-HDTV, progressive-scan signals (480p) to your HDTV, but the hardware (at least in the latest versions) is lacking. If you have an older GameCube (built before May of 2004), you can take advantage of this progressive-scan signal by using component-video cables.

If you want surround sound with GameCube games, set your A/V receiver to *Dolby Pro Logic II.* GameCube doesn't have a digital audio connection to hook up your surround-sound decoder, so there's no support for Dolby Digital surround sound.

Moving in space with Wii

Nintendo is moving ahead with its own new seventh-generation gaming console, the Nintendo Wii. (The name is pronounced *we,* or as we like to say, *wheeeeeeeee!*) Although final release dates and specifications are still not officially released, we expect that you'll be able to buy a Wii by the time you read this (the end of 2006) for no more than $250 in the United States.

Hardware-wise, Wii isn't going to be the huge leap forward over its predecessor that Xbox 360 and PS3 are over theirs. In fact, Wii will *not* support HDTV resolutions or Dolby Digital surround sound, and it will be limited to 480p resolutions (using component-video connections) and Dolby Pro Logic surround sound.

So what's so cool about Wii? Well the price is right, especially for the target audience of smaller kids, and Wii itself is tiny (only as big as three DVD cases stacked together). But the really neat feature of Wii will be its wireless controller, which looks more like an HDTV remote control than one of those thumb-muscle-building controllers that the other consoles use.

This new controller will integrate *accelerometers* that can track how the remote is moving in space — in three dimensions. So your control of movements on-screen and in your game will be less based upon how quickly you can memorize which button to press and more on how you naturally move your hand to control something. Because of this feature, Wii is expected to be easier to use for the whole family. We expect parents will *love* this when they're playing along with their kids. After all, who wants to get beaten by their 4 year old?

Connecting to Your HDTV

When you connect a gaming console to your HDTV, it's decision time. Out of the box, each of these consoles usually has only

- A composite-video connection (the yellow connector)
- A pair of analog audio cables (the red and white connectors)

If you're buying one of the many, many different bundle packages, you might get some different cables in your box, particularly with Xbox 360 and PS3.

Video connections

Most video games come with a yellow composite-video connector.

If you want to get the most from your gaming console and your HDTV, consider buying a higher-quality S-video or component-video connection — or better yet, if you have a PS3 with an HDMI connection — in either of these ways:

- Directly from the manufacturer of your console
- From a third-party cable vendor like Monster Cable (which sells a wide range of video-console cables) at www.monstercable.com

Going with composite

A composite-video connector provides the standard, out-of-the-box connection for any of these gaming consoles. A composite-video connection is also the bottom of the video-connection hierarchy; it gives you a lower-quality picture than either an S-video or a component-video connection.

If you aren't playing any games that are specifically designed for HDTVs (that is, no HDTV games on your Xbox and no 480p-resolution games for any console), this connection usually is adequate (though not optimal). But we recommend you consider a better connection.

An S-video connection can provide a significant picture boost over composite video for your console, but if you ever plan on playing progressive-scan or HDTV games, you should definitely skip over both the composite-video and S-video connections and move on up to a component-video connection.

If you use the composite-video connection (maybe you're just a casual gamer, or the console is just for the kids and they move it around a lot and can't handle the complicated cabling), you can connect several ways. All these methods work equally well:

- You can run a cable directly into one of the composite-video inputs on the back of the HDTV set.
- You can use the front-panel input (if your TV has one).
- You can connect through your A/V receiver (if you're using one) and use its video-switching facilities.

If the game is for the kids and they might either move it to the bedroom TV or take it to a friend's house, use the *front-panel* connections on your TV — just to keep the kids from messing around with any connections on the back of your system.

Stepping up to S-video

S-video is a big step up over composite video because it separates the *luminance* and the *chrominance* (the brightness and the color information in the video signal) onto their own conductors instead of trying to cram all that information into one signal onto one conductor and then separate it when it's inside the TV.

If you aren't playing 480p or high-def games on your console, you can probably just use S-video to connect your console to the TV. A definite convenience factor is at work here — your HDTV is bound to have more S-video connections available than it does component-video connections. In fact, most HDTV front-panel connections include an S-video connector, so you can easily plug and unplug the console whenever you want to move it around.

S-video connectors are the biggest pain-in-the-you-know-what in the audio/video world: They're prone to bent pins. If you let the kids make this connection, make sure you teach them how to line things up correctly before making the connection.

Depending on your equipment, you can connect S-video the same three ways as composite video: direct to the back of the TV, to the front, or through an A/V receiver. Just pick the method that's most convenient for you.

Going component

When you really want the best for your HDTV gaming experience, use component-video cables. They give you progressive-scan video for games that support it, instead of just relying on your HDTV's internal scaler system to convert the game video to progressive scan. Component-video cables also give you higher-resolution video (at least on Xbox) for true HDTV gaming.

If your console supports it (for example, if you have a PS3 or an Xbox 360) and you want widescreen, high-definition, progressive-scan video from your console, you *must* use component-video cables (or in the case of PS3, HDMI).

There's nothing complicated about using component-video cables with your console — just buy the appropriate cable system and plug it in. Keep in mind a couple of things:

- Most HDTVs don't have component-video connections on the front panel, so you must connect to the rear of the TV.
- Most HDTVs have few component-video connections (often only one set), so you might have to decide whether to connect your game, cable box, or DVD player to your HDTV.

  If you don't have enough component-video inputs on your TV, you can route your game console through an A/V receiver's component-video switching system. We discuss this in Chapter 4.

If you run component video through your A/V receiver and you want to get progressive scan (480p), make sure that your receiver's video switching has at least 10 MHz of bandwidth; to get HDTV, you need at least 30 MHz of bandwidth. You can find out how much bandwidth your video switch has by reading your A/V receiver's manual.

When you get your component-video connection set up, don't throw away that composite-video connector that came with your console. We've heard that some combinations of HDTV and consoles play through component-video cables (and work well!) but require a composite-video cable for some setup work in the console's built-in menus and controls.

Going with HDMI

If you have a Sony PS3 with an HDMI connection (the Premium model), you probably want to use that connector to hook up to your HDTV. The HDMI connection on PS3 provides a 720p, 1080i, or 1080p signal to your HDTV, depending upon which inputs you select and which your HDTV can support. This HDMI connection can also carry digital audio to your HDTV, though you might desire to use the optical digital audio output (the Toslink output) to send audio directly to your AV receiver.

Most HDTVs have a maximum of two HDMI inputs. If you're already using one of these for a disc player (like an up-converting DVD player) and another for your broadcast TV source (like an HDTV cable set-top box), you might not have room for PS3's HDMI output. You can certainly just use the component-video connection on PS3, or you can try running your HDMI through an external switching device. These devices can be built into an AV receiver (just as AV receivers can switch component, composite, and S-video) or in an external device like Gefen HDMI switchers.

Dealing with audio

Your game system probably has two connectors (red and white) for analog stereo sound. If your system has digital surround sound, you might be able to connect it to your home-theater audio system for wall-to-wall sound.

Analog audio

With analog audio cables, you can get some nice stereo sound from your HDTV or A/V receiver. You can even — if your A/V receiver supports Dolby Pro Logic II — get some *synthesized* (that is to say, fake) surround sound.

Analog audio cables don't have the true surround sound that's encoded in an increasing number of games. So you miss out on some fun audio effects (like the sounds of the bad guys creeping up behind you!).

Digital audio

To get true surround sound from a PS2/PS3 or Xbox/Xbox 360 console, you must upgrade to a digital audio cable. This connection is made using a Toslink optical cable, which is often sold as part of a package with the component-video cables.

GameCube doesn't offer an option for digital audio, though one is expected with the Nintendo Wii (wheeeee!).

Usually, you connect this Toslink cable to one of the "digital audio in" connectors on the back of an A/V receiver. Very few HDTVs accept this connection directly to their built-in audio systems.

For real surround sound, we highly recommend using a separate A/V receiver and speaker system. This approach gives you use of your five front, center, and rear speakers, plus a subwoofer — so you can really *feel* those explosions rock the room.

Chapter 15
Camcorders

In This Chapter

- Connecting your camcorder to your HDTV
- Getting into standard-definition camcorders
- Checking out HD camcorders
- Getting high-definition video from your camcorder to your HDTV

Most people make the big camcorder purchase decision when they have their first child. There's some sort of parental angst about not having home movies of your kid's first views of the world.

Well don't look now, but for people with an HDTV, the ante was just upped, as the first generations of HDTV camcorders are starting to hit the market. That's right, now you can start capturing anything you want live and in vivid HDTV color. (Although your guests might be bored of the home movies, at least they'll appreciate that wonderful high-definition clarity!)

Going HDTV for a camcorder costs you. At the time of this writing, the least-expensive HD-capable camcorders have list prices around $800 (usually nearly twice that), and not too many models are on the marketplace yet. Still, we expect the natural progression of HDTV into camcorders to continue; soon a lot more choices will be available in the marketplace, and prices will drop significantly.

In this chapter, we explore how standard-definition (SD) camcorders fit into your HDTV system, and then we go into what distinguishes an HDTV camcorder from this standard off-the-shelf fare, discussing the main competing technologies within the HD camcorder market.

Your Run-of-the-Mill SD Camcorder

Most of the concepts that we cover elsewhere in this book — in terms of SD resolution, viewing SD images on your HDTV, connecting SD devices to your HDTV, and the like — also apply to SD camcorders. Whether you use 8 millimeter, VHS, DV, or some other analog or digital format, you can view it on your HDTV.

Peering through the SD camcorder viewfinder

Many types of standard-definition camcorders are on the market at all sorts of prices and in all sorts of sizes and shapes. You can easily buy a serviceable SD camcorder for a couple hundred bucks, or you can spend well north of $1,000 for a semi-professional (or *prosumer*) model that would work well for your next hit independent movie (though most folks in this category have begun jumping to the HD camcorders we discuss later in the chapter).

Generally speaking, camcorders are categorized by two major characteristics: the system (or *format*) used to actually capture and record video, and the media — like magnetic tape — upon which the video is stored.

We're talking about the modern *digital* camcorder formats here. Some older *analog* camcorder systems are still floating around out there that record to VHS or 8-millimeter tape formats, but they're getting pretty rare.

The most common camcorder video formats are as follows:

- **MiniDV:** The majority of portable consumer camcorders use this format. MiniDV uses a small magnetic tape (considerably smaller than a deck of cards) that can record 60 or more minutes of video, depending on the quality of video you've selected, in a digitally encoded file format called just *DV.* The Mini part refers to the smaller physical tape size; other sizes of tapes can record in the DV format. MiniDV camcorders are often the easiest to deal with in the PC world because both Macs and Windows XP PCs come with free software (iMovie and Windows Movie Maker, respectively) that can import and edit video in the DV format. MiniDV can provide full SDTV resolutions, which is something older analog formats can't do.
- **Digital8:** Camcorders that use the Digital8 format are a slightly cheaper alternative to MiniDV. These camcorders are bigger because they're built around the physically larger *Hi8* 8-millimeter tape format. Like MiniDV, Digital8 records in the DV file format, so it can offer a good quality SDTV picture.
- **DVD camcorders:** A growing number of camcorders record their video directly to a small, but otherwise standard, DVD-R or DVD-RAM disc. These camcorders record directly onto the disc by using the MPEG-2 compression system (the same system used by regular DVDs), so you can just pop the disc out of the camcorder and into your DVD player for watching on your HDTV screen (very handy!). The disadvantage here is that it's much more difficult to edit your video on your PC. For most people, unedited video isn't always all that great. We tend to discard huge chunks of footage of our feet or the wall or other unintended bits around the video we're really interested in.

- **Flash memory camcorders:** Camcorders that record directly onto solid-state *flash* memory (the same sort of memory used by digital cameras) are the newest category. These camcorders usually record video by using MPEG-2 compression, and they can offer a good quality SDTV picture. The biggest advantage of these recorders is that they simply do away with the tape, so they're smaller and they deliver you from the physical shortcomings of magnetic tapes with their rewinding and fast forwarding. On the downside, however, these camcorders usually have shorter recording times (watch out if the kid's recital is too long), and you can't easily edit the MPEG-2 video on the computer.

  Some camcorders substitute a built-in (and iPod-sized tiny!) hard drive for the flash memory.

Beyond the format of the camcorder, a few other major areas separate the performance of a camcorder, including the following:

- **The resolution of the CCD:** The CCD (or *charge-coupled device*) is the *image sensor* — "retina" of the camera — that picks up the image sent through the lens and converts it to an electronic signal. Different CCDs have different resolutions. Generally speaking, the higher the *effective* resolution (meaning the number of pixels on the CCD that actually pick up light from the lens — sometimes not all of the pixels are used), the better.

  Effective resolution comes into play most often when you're dealing with a widescreen image. Sometimes the CCD itself is physically shaped for a 4:3 image, so only part of the sensor is used for the widescreen 16:9 image. The *total* resolution of that sensor would therefore be higher than the *effective* resolution that is actually picking up a picture.

  Not all camcorders use CCD image sensors. Some use *CMOS* (complementary metal-oxide semiconductor) instead. Regardless of sensor type, the observations we make here about CCD apply.

- **The size of the CCD:** The trend for miniaturization is in full effect in the camcorder world, where every single component inside the camera is shrunken to make the whole package smaller. This is true of the CCDs themselves, but unfortunately, shrinking the CCD isn't always a good thing. In fact, for the same number of pixels, a smaller CCD usually isn't as good at picking up a noise-free (distortion-free), richly colored image. It's just physics at work when you cram all those pixels so close to each other. The least expensive camcorders have ⅙ CCDs, whereas others have ¼ or ⅓ CCDs. If you're serious about your video, you probably want at least ¼ CCDs.

- **The number of CCDs:** Most cameras use a single CCD to capture the whole image, but a number of fancier models have begun using three — one each for red, green, and blue colors. This three-CCD approach provides truer color reproduction.

- **Widescreen mode:** Most new camcorders can record standard-definition video in a widescreen (16:9) format. We highly recommend this feature if you want to play the video on your HDTV (which, of course, is widescreen itself). Down with the black bars!
- **The audio recording system:** Different camcorder formats and even different camcorder models within a format record audio in different ways. Generally speaking, the greater the number of *bits* used for audio (for example, 12-bit audio versus 16-bit audio), the higher the quality of the sound you pick up. Sixteen-bit audio is CD quality.
- **The optics:** Simply put, a better lens gives you a better picture. For example, a better lens might be better in low-light situations, such as pretty much *all* indoor video shoots. Many of the qualities of a lens are somewhat subjective, but here's one objective to keep your eye on: the optical zoom the lens provides. Like digital still cameras, digital camcorders can zoom the picture by using the lens itself (the *optical zoom*) or by electronically "blowing up" a smaller section of the digital image picked up by the sensor (the *digital zoom*). Optical zoom preserves the full resolution of the picture; digital zoom reduces it and gives you a blurrier image. If you're going to be shooting a lot of video from longer distances, go for greater optical zoom. For a camcorder, look for an optical zoom of 10x or greater.
- **Other features:** After you get past the preceding criteria, you get into a host of other features and functionalities that vary among different individual models. Things like the ability to take still images, battery life, size and shape of the camcorder, ergonomics (the ability to easily use all of the controls), accessories (like external microphones), and dozens more features all can impact your decision of a particular camcorder. This is where you need to do your homework; try out different models at the store and see what works best for you.

Our favorite sites for reading well-done reviews and getting good buying advice for a camcorder are CNET (`www.cnet.com`) and Camcorderinfo.com (`www.camcorderinfo.com`). We highly recommend you peruse both of these sites before you buy.

Connecting your SD camcorder to your HDTV

Chapter 3 gives you an idea about a concept we strongly believe in: a *hierarchy* among the many connection options available on most video sources (including camcorders). That is to say, these connections have a definite good-better-best rank, and here they are, starting with the best:

- **Going digital:** Most camcorders have a FireWire (1394) output, and many models have DVI or HDMI outputs. If your HDTV has a FireWire input, you might be able to use the FireWire connection on your camcorder (found on most MiniDV camcorders) to connect digitally. If this works with your camcorder (and it doesn't always work), it produces the best-quality picture on your HDTV. We expect digital connections (FireWire, DVI, or HDMI) to become available on more and more camcorders.

Even if both your HDTV and your camcorder have FireWire connections, you might not be able to use them to connect the two together. Check your camcorder's manual — many times you can use only FireWire for connecting to a PC.

You *can* use the FireWire connection to send your video to your PC, edit it (using iMovie on a Mac, or Windows Movie Maker on Windows XP), and then burn your own DVD or copy to your D-VHS recorder. (See Chapter 13 for more about D-VHS recorders.) This gives you a very high-quality video source for HDTV based on your home movies.

- **Going with S-video:** Most modern camcorders (particularly MiniDV models) include an S-video output. The S-video connection provides a better picture by separating the color and brightness portions of your video onto separate cables — allowing your HDTV to display them without having to use the *comb filter* to separate these picture elements.
- **Using the composite solution:** The least-attractive solution for connecting your camcorder to your HDTV is to use the yellow composite-video connection. We always recommend using S-video over composite video, but some older camcorders simply don't give you that option.

Some day in the future, we expect that non-HDTV camcorders will also offer the *component*-video cable connection method — this will make sense as more camcorders become capable of dealing with progressive-scan video. Right now, only HDTV camcorders can use this connection method. In terms of the hierarchy, we place component video below the digital connection methods (like HDMI), and above S-video.

Why the wait for HDTV camcorders?

If you've been waiting for HDTV to come to your camcorder, it's only been recently that this has been made possible, for largely two reasons:

- The digital signal processors (DSPs) have become small and smart enough to handle the immense volume of data created by the high-definition CCD imaging sensors in your camcorder. Each frame offers a megapixel of resolution or more, and this has to be processed and recorded in real time to digital video tape.
- Consumers have just started caring! As the 16:9 television sets replace the 4:3 sets, people can see HDTV in its native mode, and that means a bigger market for the camcorder manufacturers to sell to.

So while you might not have known it, merely buying this book is helping get more HDTV camcorders to market! Thanks for helping out.

Enter HD Camcorders

So if you can link your SD camcorder to your regular TV, what's the big deal with HD versions? Lots of pixels, for one thing — three times as many pixels as the best offered by NTSC versions, encoded as a standard MPEG-2 stream.

Ever wonder why the SD camcorders look so poor on your HDTV? It's simply a reflection of the lower resolution of the SD stream when scaled up and viewed on your HD screen — the two were simply not meant for each other.

But lots of pixels mean lots of megabytes, too. An uncompressed 1,280 x 720 BMP file can be almost 3MB in size — that's just a single frame of video ($\frac{1}{30}$ or $\frac{1}{60}$ of a second's worth of video). That's the price of high-definition.

Of course, HD camcorders use a ton of computer horsepower to *compress* these video frames so they use less storage space — but HDTV video still uses a ton!

High-def camcorders have two main competing standards:

- **HDV:** *HDV* (high-definition video) camcorders use the same DV format as MiniDV standard-definition camcorders, but the DV *encoding* format used by those camcorders is thrown away and replaced by good old MPEG-2. MPEG-2 allows the camcorder to put a lot more data on the same tape, and it provides support for both 720p and 1080i HDTV formats. Most HDV camcorders record in one format or the other, however, so when you shop for an HDV camcorder, you must choose 720p or 1080i. Note that some HDV camcorders can forego the tape and record directly to a hard drive. The biggest advantage of HDV (besides the fact that it uses the tried and true MiniDV tape format) is that you can easily edit the video it produces on PCs. Many Windows programs can support HDV editing, and all Macs with Apple's iMovie can edit HDV out of the box.
- **AVCHD:** A newer format for HD camcorders is *AVCHD* (or Advanced Video Codec High Definition). AVCHD uses the MPEG-4 encoding system, which is more efficient than MPEG-2 and is often called MPEG-4 AVC, hence the AVCHD name. AVCHD camcorders typically use small recordable DVDs as their storage medium, just like DVD standard-definition camcorders. However, they can also use internal hard drives or flash memory systems to record their 720p or 1080i video. AVCHD camcorders are usually nice and compact because they don't need the bulky tape-handling mechanism, but they typically can't hold quite as much video as a tape-based machine, and the MPEG-4 video (plus the DVD format) isn't as easily transferred to and edited on a computer.

You can play the DVDs recorded by AVCHD camcorders in a Blu-ray disc player (or Sony PlayStation 3, which has a built-in Blu-ray disc player). You can read more about Blu-ray in Chapter 11. Note that these discs won't play in an HD DVD player.

You can also find some *proprietary* HDTV camcorders, which use neither of these formats but instead use some vendor-specific system. For example, Sanyo's Xacti VPC-HD1 camcorder records 720p video by using MPEG-4 compression and saving the video onto flash memory, but it doesn't use AVCHD. Rather, it uses Sanyo's own system for such recordings.

When dealing with the higher resolution and more detailed image of a high-definition camcorder movie and its display, everything is more noticeable. Errors such as shaky handheld shots, too much panning, and zooming too fast can make people dizzy (if not nauseated) when viewed on large display. Think about using a tripod or bracing yourself more securely when making videos.

Other than the resolution, much of the camcorder will look familiar to you — expect the same high-speed interfaces (such as FireWire) and even the same tape formats (in the case of HDV).

Checking out the HDV camcorders

Although some specialty camcorders have offered HD for some time (see the sidebar, "Live from outer space . . . in HDTV"), the first consumer-oriented and -priced product came from JVC: the GR-HD1 camcorder, which retails around $3,500 but can be had on the street (or Internet) for $2,700 or less. Figure 15-1 shows this HDTV beauty.

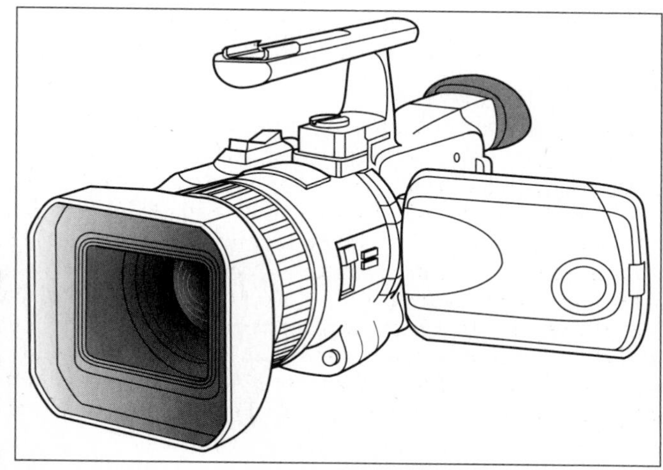

Figure 15-1: Record in high-def with the GR-HD1.

The GR-HD1 camcorder gives you a way to get your personally recorded 1,280-x-720-pixel content to your HDTV system and even archived on your D-VHS VCRs (discussed in Chapter 13), which are also sold by JVC and can play prerecorded D-Theater (JVC's brand name of D-VHS) HDTV movies. JVC includes a FireWire interface and software with the camcorder that converts 720-lines-of-resolution, progressive-scan HD footage (720p) into progressive-scan, anamorphic widescreen DVDs by using your PC — that's cool! Material recorded either in DV or HD format can be down-converted or up-converted as needed for playback (depending upon what format your HDTV requires).

The GR-HD1 isn't an HDV camcorder, but actually uses a proprietary JVC variant of this system known as *JVC ProHD*.

The GR-HD1 has been joined in the marketplace by a number of new HD camcorders, and that number seems to be getting higher every day. Some of the more interesting models available include

- **Sanyo's Xacti VPC-HD1:** As we write in mid-2006, Sanyo has the cheapest HD-capable camcorder on the market, the HD1. This model can support 720p video. The actual resolution is 1,280 x 760 pixels. Like a growing number of video cameras, the HD1 does away with the tape and uses a solid-state memory card (like a digital camera uses) to record video. You can record up to 41 minutes of high-def video on this model by using a 2GB *SD Card* memory card. With a list price of $799.99, this HD-capable camcorder is the cheapest yet.
- **Sony's HDR-FX1:** Sony's entry into the prosumer HD camcorder market is a bit more expensive (okay, a *big bit*). It lists at $4,999.99, but is available online for closer to $3,200. This killer machine can record video at the full 1080i resolution and features a lens, three CCDs (explained earlier, in "Peering through the SD camcorder viewfinder") for better color reproduction, and more. It's an awesome camcorder if you can swing the asking price. The HDR-FX1 uses the HDV video format.

  Sony also has a cheaper HDV model (with fewer bells and whistles, of course), the HDR-HC3, which retails for under $1,800.

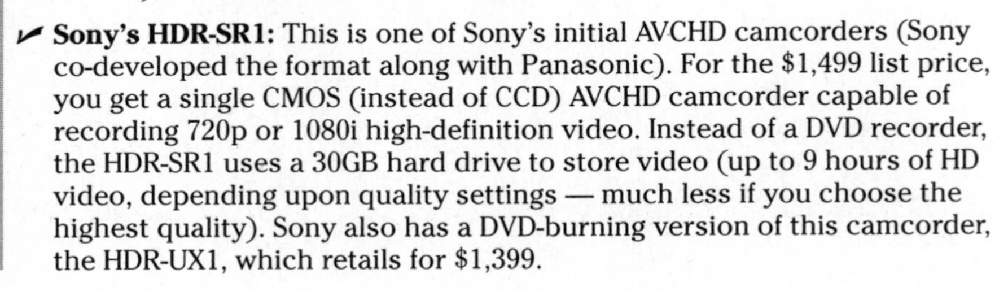

- **Sony's HDR-SR1:** This is one of Sony's initial AVCHD camcorders (Sony co-developed the format along with Panasonic). For the $1,499 list price, you get a single CMOS (instead of CCD) AVCHD camcorder capable of recording 720p or 1080i high-definition video. Instead of a DVD recorder, the HDR-SR1 uses a 30GB hard drive to store video (up to 9 hours of HD video, depending upon quality settings — much less if you choose the highest quality). Sony also has a DVD-burning version of this camcorder, the HDR-UX1, which retails for $1,399.

The other developer of AVCHD, Panasonic, has begun showing prototypes of its AVCHD camcorders, but no final pricing and availability was available as we wrote. Expect to see several models from Panasonic on the market by the time you read this.

You might not *always* want to record in high definition, particularly if you're making regular standard-definition DVDs of your footage for your non-HDTV-owning family and friends. (By the way, it's time to evangelize and get them on board the HDTV train!) When you're evaluating an HD camcorder, be sure to look into its support for standard-definition video, as well. All the models we discuss can also shoot standard-definition video.

If you want to check out the most recent HDTV camcorders on the market, see our site at `www.digitaldummies.com` or go to `www.camcorderinfo.com` for great reviews of the latest HDTV camcorders. And if you want to know more about digital camcorders and how to make movies, check out *Digital Video For Dummies,* by our fellow *For Dummies* author Keith Underdahl (Wiley).

Live from outer space . . . in HDTV

You're probably used to having ultra-nice pictures from space of Earth in all its glory, but we bet you'd never guess that HDTV was involved. That's right, even the space-shuttle crew is into HDTV camcorders. HDTV equipment flew as early as STS-95 (1998) and included a Sony HDW-700A high-definition television camcorder, wide-angle lens, battery packs, and video-recording tapes.

NASA is using the high-resolution images to provide clearer pictures about life on the space station and to improve the documentation of space exploration.

The system is enhancing the capability of NASA scientists, researchers, and engineers to conduct their research, monitor experiments, and record the data visually. HDTV also allows the public to experience NASA's explorations more realistically by making the footage available over NASA TV.

So here's a tip from those who learned the hard way: Next time you're in space, watch out for space radiation. It can cause degradation in your CCD image sensor (the silicon chip inside the camera which contains a rectangular array of light-sensitive cells). The degradation showed up on NASA pictures early on, as a loss of several pixels on images taken on board. The camera lost between 5 and 15 pixels per day. So NASA created a self-correcting camera that replaces the bad pixels with an average of the luminance and chroma from adjacent pixels. Problem solved. (That's your tax dollars at work!)

Getting HD video from camcorder to TV

The biggest issue for many folks interested in an HD camcorder is figuring out how to get that nice HD video from the camcorder to the TV. In the SD world, it's quite easy to create a DVD on your PC that plays your videos for you. Tools like Windows Movie Maker and iMovie allow you to create professional-looking DVDs with menus and soundtracks and all the niceties.

Unfortunately, as we write, there isn't an easy solution for editing HD video and then recording it onto a disc. (We expect this will change as HD DVD and Blu-ray recorders become more common in desktop and laptop PCs.) This lack doesn't mean you can't get your HD footage onto your TV however. What's the point otherwise, right? You can take a couple of approaches:

- **Use the HD camcorder as a playback device:** You can simply plug the HD camcorder directly into your HDTV to play your videos. Most HD camcorders have component-video outputs, and a growing number have HDMI outputs. In these cases, you simply plug this output (and its corresponding audio output in the case of component video) into your HDTV and you're ready to go. Set the camcorder to play mode and enjoy the precious memories!
- **Use a PC as a playback device:** If you're transferring your video to a PC for storage and editing, you can take the resulting movie file and use a PC connected to your TV for playback. See Chapter 16 for more information on hooking up a PC to your HDTV.
- **Use the optical disc you recorded on:** The AVCHD camcorders that record to recordable DVD media are perhaps the simplest in terms of big-screen playback — if you have a Blu-ray disc player that is. Just pop the disc out of the player and throw it into your Blu-ray disc player or Sony PlayStation 3 and you're set. Easy as pie. Of course, only a handful of folks have such players right now, but over time we expect them to become quite popular (due to the expected sales of the PlayStation 3).
- **Burn your own high-definition disc:** This is in the realm of "coming soon" as we write, but both Blu-ray and HD DVD recorders are starting to appear on a handful of PCs. When the inexorable march of time and technology brings recordable Blu-ray and HD DVD to the masses, *and when* PC and Mac disc-authoring software packages support it, you'll be able to capture your high-definition video from camcorder to PC, edit it, and then burn your own HD discs. We, for one (or is that, "for two"?), can't wait!

Chapter 16
Gadgets

In This Chapter

- Getting a PC into your HDTV
- Going with a jukebox
- Serving up media
- Finding new gadgets

We love gadgets. If you're getting into HDTV, we expect you're no different from us (or the rest of the world).

In this chapter, we cover cool HDTV gadgets that can act as direct sources of content to enhance your video oasis. Whether it's a high-powered home-theater PC or specialty gear destined to connect to your HDTV, you can really boost your HDTV's usage with just the right accenting gear.

We also talk about HDTV image scalers, which are focused on improving *all* the content going into your HDTV set. All HDTVs have some sort of image-scaling system built into them, but cost and design constraints often keep them from being as good as they can be. When you get into the world of stand-alone scalers, the sky is truly the limit.

Finally, we talk about really high-end gadgets that can store all of your media digitally and serve it up across a home network to any HDTV in the home!

HDTV technology is only slowly starting to infiltrate the gadgets that you'd use with your HDTV system — in other words, most of the stuff in this chapter doesn't send an HDTV signal to your HDTV. Only a few true HDTV gadgets are out there yet, but we tell you what's available today and where to find it.

Home Theater PCs

Although nearly all of today's PCs are *multimedia*-capable (they can display pictures, play sounds, and show video), a select few PCs can be considered *home theater PCs* (HTPCs) that can feed video and surround-sound audio into an HDTV.

HTPCs are simply high-powered PCs running the Windows, Linux, or Mac operating system (the majority use Windows), that have been specially configured with hardware and software that lets them operate as the DVD player, TV tuner, or even DVR (see Chapter 12) source for your HDTV.

What's inside an HTPC?

No rigid set of rules defines what makes a regular PC into an HTPC, but here's what we recommend:

- **Video card:** Perhaps the most important item in an HTPC (particularly one that feeds video into an HDTV) is the video card. This specialized set of computer chips (the *GPU* or Graphics Processing Unit) spares the computer's CPU from most of the heavy lifting of video processing. Both ATI's Radeon series (`www.ati.com`) and NVIDIA's GeForce series (`www.nvidia.com`) include high-end video cards and GPUs that can support HDTV resolutions and support HTPC applications.

  Many manufacturers build specialized video cards based upon NVIDIA's and ATI's technologies. So you might, for example, end up purchasing a card from a company like ASUS, Gigabyte Technologies, or MSI for your HTPC and actually be running an NVIDIA GPU with some additional hardware and software provided by the company who manufactured the card you're using.

  A growing number of video cards include HDMI connections for a digital connection to your HDTV. If your HDTV is equipped with HDMI, you might want to consider a card such as ASUS EN7600GT/HDTI/256MB (`www.asus.com`), which includes this connection.

- **Audio card:** If you want to support the surround sound (Dolby Digital and other systems like DTS) found in HDTV broadcasts and on DVDs, you need a relatively high-end audio card in your PC. A card like the Sound Blaster xFi (`www.soundblaster.com`) is a good choice.

- **CPU:** There isn't a hard-and-fast rule here, but you build an HTPC around a PC with a fast processor. Look for either

  - A 2.8 GHz or faster Pentium 4 processor

    Microsoft requires a 3 GHz Pentium 4 for full 1080i HDTV playback with Windows Media.

  - An AMD Athlon 64 3200+ processor

- **Hard drive:** You need a decent-size hard drive on any media-centric PC. Media take up room; you need enough to store such data as MP3 and other music files.

  If you want an HTPC as a DVR feeding your HDTV, go for

  - At least 250GB of hard drive space
  - Plenty of FireWire or USB 2.0 jacks for attaching additional hard drives

- **TV tuner:** If you want to use the HTPC as a DVR, a TV tuner card (also called a *capture card*) is essential. Most of the DVR hardware/software kits on the market include a TV tuner. Because we're talking about HTPCs in the context of an HDTV, you should look for a system that can capture ATSC high-definition channels as well as standard definition.

  If you just want the PC for playing DVDs or Internet content, a TV tuner card inside your HTPC might not be important.

  You don't need to put a TV tuner card *inside* of your HTPC. Many manufacturers make inexpensive ($100 to $150) external USB TV tuners that can plug into a USB 2.0 port on your HTPC.

- **Software:** You might eventually want a wide range of software for your HTPC, such as software that turns your PC into a PVR or organizes your media library.

  The most important software might be a video-utility application, such as PowerStrip (`http://entechtaiwan.net/util/ps.shtm`), which helps you perfectly match the resolution of your video card output to your HDTV's display resolution.

A great place to get advice, see the results of people's projects, and generally dig into the topic of HTPCs is the AVS Forum Web site (`www.avsforum.com`). Check out the section titled "Home Theater PCs" for all the info you could ever want.

Building or buying an HTPC

You can buy an off-the-shelf HTPC from most major PC vendors and many smaller and HTPC-focused vendors like Alienware — not so small, as it is now part of Dell — or Nexus in the Netherlands (`www.nexustek.nl`). Some of these manufacturers have gone to great lengths to design HTPCs that fit right into a home-theater environment — they look good next to your shiny new HDTV, they have all the right ports for expansion and connectivity, and (most importantly) they run quietly.

Can I use a PC for Blu-ray or HD DVD?

PC manufacturers are starting to equip their high-end PCs with HD DVD and Blu-ray disc drives. This makes a lot of sense simply because these disc formats can hold a lot more data than standard DVDs can (ten times as much, or more, depending on the format). So you might guess that using an HTPC with a Blu-ray or HD DVD drive would be an easy way to bring these high-definition video formats to your big-screen HDTV.

Well, that's actually an iffy thing. Both HD DVD and Blu-ray are built with a very strong copy-protection system called *HDCP* (which we discuss in Chapter 3). This system is designed to prevent unauthorized copying and distribution of movies by limiting the devices that can digitally connect to your HD DVD or Blu-ray disc player. When the player is inside your PC, the only way you can use an external monitor or HDTV to view a movie in its full high-definition resolution is to use an HDMI (or DVI) connector that supports HDCP.

Seems pretty simple, but in fact, very few PCs have video cards with HDMI connectors, and even fewer have HDMI connectors with HDCP support. So in order to watch a high-definition movie, you have to jump through some pretty major hoops right up front. And if you pass these initial tests (HDMI connectors and HDCP support), you're still not guaranteed that everything will work as advertised. We've read many reviews and tests of some of the initial PCs so outfitted, and the consensus seems to be that Blu-ray or HD DVD playback on an HDTV *might* work.

Unfortunately, in many cases two issues seem to crop up. First, HDMI and HDCP implementations might differ between the PC and the HDTV, keeping the HDCP system from making a proper "handshake" that would authorize the system to play a movie. Second, the HDCP systems on the PC just don't have enough computing horsepower to keep up. One review we read stated that the PC's CPU was basically maxing out trying to keep up with the HDCP decryption, and it was causing the movie to break up in a very distracting fashion.

Our advice is not to give up on HTPCs and Blu-ray and HD DVD, but also not to expect this combination to be better than a standalone player. If your HTPC is a laptop, you won't have any problems viewing these movies on the laptop screen — and that alone might make any extra expense of an HD DVD or Blu-ray drive worthwhile.

But you don't need to buy a premade HTPC if you don't want to. As long as you have a PC that meets the minimum hardware requirements, you can add software and hardware to turn it into an HTPC. In the following sections, we talk about the major software solutions that make a PC into an HTPC.

Media Center Edition 2005

An easy way to get into the HTPC game is to buy a new *Media Center PC* with the Microsoft Windows XP Media Center Edition 2005 (MCE for short) operating system. Media Center PCs meet stringent minimum hardware requirements that allow them to work as HTPCs right out of the box. The big advantage of an MCE PC (besides having the hardware checklist all crossed off) is the software: The Media Center software provides a nicely integrated experience for such home-theater functions as DVD playback, DVR functionality, and TV-watching.

Media Center Edition uses a unique *ten-foot* interface (one that you can manipulate from about ten feet away on the sofa, instead of two feet away at a desk) that hides the normal Windows interface behind a full-screen application (Microsoft calls it an *environment*) and lets you access and playback all of your PC media — TV, DVDs, music, and pictures — with a remote control.

Media Center PCs work great when connected directly to your HDTV, but they can also feed high-definition content to your HDTV from anywhere in the house if you have an Xbox 360 (see Chapter 14). The Xbox 360 comes with software that lets you remotely access all of your Media Center Edition content across a wired or wireless home network. One thing to keep in mind is that HDTV content is going to work reliably only across a wired (Ethernet) network — a wireless network typically won't have enough bandwidth to handle HDTV from your Media Center PC.

Getting a view of Vista

Some time in 2007 (and we say "some time" because the actual date has been pushed back a few times), Microsoft will replace Windows XP with a new operating system called Vista. In fact, Microsoft will replace the three versions of XP (Home, Professional, and Media Center Edition) with *six* versions of the new OS. Two versions — Vista Home Premium and Vista Ultimate — will include the features and functionality that are now included in Media Center Edition. Included in these versions will be a truly important new feature for HTPC users who rely upon cable TV: support for the CableCARD systems (discussed in Chapter 8). This will allow you to lease a CableCARD from your cable provider, plug it into your HTPC, and pick up "premium" cable HDTV channels on the PC, which is something you need to use an external cable box for today. We're looking forward to this hitting the streets — though we must warn you that cable companies to date have only reluctantly and lukewarmly supported CableCARD. Hopefully Microsoft adopting CableCARD will help push the system forward towards broader acceptance.

Understanding Viiv

The folks at Intel have designed a new PC architecture called Viiv, which is designed from the ground up to be the basis for HTPCs and entertainment-centric PCs in general. Viiv includes processors, GPUs, and networking hardware and wraps this hardware together with some software enhancements that go beyond a typical PC — including but not limited to Media Center Edition.

Perhaps the neatest feature of Viiv-powered PCs is the fact that they're designed to operate more like a piece of consumer electronics gear (like a DVD player) than a PC. That means you can turn them on and off and get a nearly instantaneous response. Viiv PCs also have enhancements to their hard drive storage (incorporating RAID for redundant storage), 7.1 channel surround sound audio, and more. All in all, the Viiv logo is an easy way to be sure a PC is ready for your HDTV. For more information on Viiv as well as a list of Viiv-enabled PCs, check out the following URL: `www.intel.com/products/viiv/index.htm`.

You don't *need* Viiv to run Media Center; Viiv is just a PC architecture and software solution that makes Media Center work better.

The latest version of Media Center Edition (the elegantly named Update 1 for Windows XP Media Center Edition 2005) adds support for high-definition TV, including DVR support. Previous versions of Media Center Edition didn't support HDTV at all — at least not without add-on software.

SageTV

One of the more popular alternatives to Windows Media Center Edition 2005 is a software package called SageTV (`www.sagetv.com`). SageTV Media Center, a Windows-only package, turns your PC into a full-function DVR that can record both standard- and high-definition TV broadcasts on your PC's hard drive. SageTV, like Media Center Edition, is designed with a ten-foot interface that displays on your HDTV and lets you easily control the software with a remote control — instead of requiring you to use your mouse and keyboard.

Also like Media Center Edition, SageTV can control a TV tuner that's built into your PC (via a PC card or attached to a USB 2.0 port) or it can control an external cable set-top box or satellite receiver by using an infrared controller.

The coolest features of SageTV, in our opinion, are the home-networking abilities. The Sage folks sell inexpensive media extenders ($109 for the wired Ethernet version and $159 for the wireless version) that let you view DVR content from your SageTV-equipped PC on other TVs around the house. Keep in mind, however, that these extenders can handle only standard-definition TV — so they're great for the small TV in a bedroom, but not so great for your HDTVs. (Instead, extenders are best located directly next to your HTPC.)

Apple's Front Row

The folks at Apple Computer are about as advanced in the media business as anyone in the PC world — the iPod, iTunes, and the whole industry built around them are proof positive of this. But so far, Apple is a bit behind when it comes to the whole HTPC game. It's not that Apple computers don't make good HTPCs, but rather the Mac OS X family just doesn't have a full equivalent to Media Center Edition.

What Apple does have today is a program called Front Row, which is included in most new Macs. Front Row includes a ten-foot interface, similar to Media Center Edition, and a cool remote control that lets you view your media on the HDTV and quickly access it all. What Front Row doesn't have today is the TV-viewing and DVR functions that Media Center Edition brings to the table.

We expect that Apple will add this functionality in due time. The forthcoming iTV (a video player) might help bring this about. In the meantime, Mac users can look at third-party solutions like El Gato's EyeTV (`www.elgato.com`) to add this feature to their existing Macs.

MythTV

If your PC operating system tastes run away from the commercial (Windows) and more towards the non-commercial (Linux), MythTV might be the HTPC software for you. MythTV is an open source, free (as in you're free to download it without paying a penny), DVR media center application for Linux PCs.

MythTV is free in another way — as an open source program, it is designed to allow other programmers and developers to freely create additional features. So while the core program is really designed solely for the DVR functionality, other folks have created add-ins for MythTV that let you use the program to play and manage your digital music collection, view photos, and more, all using your HDTV and a remote control as the primary interface.

By the way, MythTV also runs on the UNIX-based Apple OS X system, on both PowerPC and Intel-based Macs. You can get all of the details about MythTV on the MythTV Wiki page at `www.mythtv.org/wiki/index.php/Main_Page`.

Video Jukeboxes

It isn't unusual to find movie junkies with thousands of DVDs. Organizing and accessing these DVDs can be troublesome at best, and the more you have, the harder it is to find one movie that you really wanted to watch. Enter video jukeboxes, gallantly riding to the rescue.

Jukebox 101

A DVD video jukebox does what you'd think — stores massive numbers of DVDs so that you can watch them on your HDTV in a more organized and accessible way, which is great when you want to watch movies all over the house.

The entry-level video jukebox is the *megachanger,* also called a DVD *carousel.* These DVD devices can store 400 or more DVDs and play any of them, usually one at a time. Some DVD changers can even control a *second* changer (doubling the capacity, for example, from 400 discs to 800)! A good example is Sony's DVP-CX995V, which retails for $399.99 and holds 400 discs. It even plays back high-quality SACD audio discs!

A DVD jukebox needs the following features (which become more and more essential as the jukeboxes get bigger):

- A library function that can keep track of what disc is in which slot
- An easy way to access that library function, such as
  - An OSD (on-screen display) that you see on your HDTV
  - An LCD screen on the jukebox itself
- An easy way to feed information into the library

  The *easiest* way to enter information is with a keyboard, *not* with the remote control. Depending on the jukebox, you connect a keyboard in one of two ways:

  - Directly to the jukebox
  - By linking to a PC (usually a USB connection)

  Check whether the device automatically looks up disc-title information on the Internet — it can save you a ton of data-entry time.

A video jukebox needs the same basic DVD player features that we cover in Chapter 11, such as

- Progressive scan
- Support for other disc formats, such as CD-R and MP3 CDs
- Support for new *audio* formats, such as SACD or DVD-A
- A full range of digital audio outputs and HDTV-friendly video outputs (such as component video)

You can find DVD jukeboxes anywhere you find regular DVD players — at standard online and brick-and-mortar electronics retailers.

Jukebox 301

For advanced movie watchers (okay, for those with more advanced budgets), you can leave behind the stand-alone DVD jukebox and move into networked DVD jukeboxes and even into hard-drive-based video jukeboxes.

These devices are really more properly classified as *media servers* — computer-based devices that store various media (such as video, audio, and photos) on a hard drive. These media servers are high-end video systems, with price tags running into the thousands of dollars, and are typically found at high-end boutique electronics and home-theater dealers.

VideoReQuest

Audio is the main focus for the folks at ReQuest (`www.request.com`), but they also have dished up a neat product for the high-end video enthusiast. The VideoReQuest combines off-the-shelf DVD jukeboxes (Sony's DVP-CX777ES 400 disc changer) with ReQuest's own hardware and software system to tie everything together and make controlling your video easier than you'd ever believe.

With the VideoReQuest, you can control up to four of these changers (that's 1,600 DVDs!) from anywhere in the home. The main VideoReQuest unit hooks up to your home network via an Ethernet connection, and any PC in your home can control it (start and stop movies, sort through your collection, and more). The coolest feature is that the VideoReQuest uses this network connection to run queries on the Internet that provide you with detailed information about the movies in your system. You can see cover art, titles, actors, directors, and more.

When watching a movie, you can sort through your titles by all of these criteria using a PC, a remote control, and an on-screen interface on your HDTV or even remotely over an Internet connection. Pretty cool stuff.

As we write, VideoReQuest doesn't include any facilities for playing high-definition HD DVD or Blu-ray discs. We expect that this feature will appear before too long, but there have been no official announcements yet.

Escient's Fireball

Escient (`www.escient.com`) manufactures a wide range of digital media server systems, mostly focused on providing a centralized, whole-home digital audio system.

But Escient has more than just music; the company has three versions of its Fireball system (the MX-111, MX-531, and MX-752, ranging in price from $1,999 to $5,999). These systems include the digital music functions of other Escient systems, but they add in the ability to control one or more DVD jukebox changers.

In many ways, the Fireball is like the VideoReQuest — it supports the same Sony DVD changer, and it offers many of the same network-based features. One feature we really like is the support for Internet radio through the Shoutcast system. Escient runs its own radio station finder service that lets you quickly find the stations you want.

Kaleidescape

The Kaleidescape System is a truly high-end media server solution (www.kaleidescape.com).

The big advantage of the Kaleidescape system is its modular, multiroom support — but you'll pay accordingly. A Kaleidescape system can *start* at over $27,000 for the base system. As we write, Kaleidescape has announced a new, cheaper, and slightly less capable system that starts off at a mere $10,000. It's just plain expensive to build hard-drive-based systems that do what Kaleidescape does, which is why many other high-end systems like VideoReQuest use DVD changers instead.

The Kaleidescape System is similar in functionality to the Media Mogul systems, only everything is cranked up a bit. In particular, the Kaleidescape is high-definition-ready and supports 720p and 1080i HDTV. This multiroom system consists of

- A central server with interchangeable hard-drive cartridges
- A movie player that connects to your HDTV

  As you expand your budget and desires, you can add movie players connected to HDTVs (or any TV) in other rooms.
- A DVD reader

MIA: The Roku HD-1000

The folks at Roku Labs (www.rokulabs.com) sell some of the neatest digital audio players around. Along with the Squeezebox (www.slimdevices.com), Roku's networked music players are our favorites. And until recently, Roku sold one of our favorite inexpensive devices for feeding digital content to your HDTV: the HD-1000 HD Photobridge. This device provided a high-definition feed over component video cables, which allowed you to display your photo albums and artwork on the big HDTV screen. It also was a networked music player that allowed you to access all of your music on the big screen.

Roku has discontinued this product but will be announcing a replacement in early 2007. We haven't seen anything about the new product, but given how much we liked the old one (Danny has one running on his HDTV and uses it often), we expect it will be worth keeping an eye on.

The whole system connects via Ethernet cabling. It can use a broadband (cable or DSL modem) connection to the Internet to update content, such as the movie-guide service, and upgrade the system's software.

Boosting Your Video Signal Quality

When you start getting into the higher reaches of HDTV, it's not just the gadgets, gizmos, and boxes that provide new and better sources of video content for your HDTV that get you excited; it's also the devices that make *any* content from any source look better on that big HDTV display. These devices — called *image scalers* or *video processors* — take standard-definition inputs, perform some internal computations, and output a higher-quality high-resolution signal for your HDTV. In other words, they convert the signal coming from a standard DVD player, VCR, or other device into something approximating a true high-definition signal.

All modern HDTVs have some sort of image scaler in them. At the very least, they have an internal device that *de-interlaces* interlaced video content (like 480i standard-definition TV and even 1080i high-definition content) to play it back on the progressive-scan display within an HDTV.

Beyond this, most of today's HDTVs — no matter the type — are *fixed-resolution* displays. They have a finite and unchanging number of pixels that physically make up the picture you're seeing (called the *native* or *display resolution*). These HDTVs use an internal scaling system to take whatever input resolution you feed them from your video sources (TV, DVD, and so on) and convert it to the display resolution. This feature allows you to watch, for example, a 1,920-x-1,080-pixel 1080i program on your 1,280-x-720-pixel HDTV display.

Check out Chapter 21 if any of these terms are unfamiliar.

So we know what you might be thinking at this point: "Um, hello guys, if my TV already has this, why would I want to spend my money on another scaler?" Well, we have two answers to that. First, you don't need to! Most folks don't, and they're still happier than pigs in mud with their HDTV sets and the pictures that they provide. The second answer is a bit fuzzier. You don't *need* one of these devices, but you might *want* one if you have a few things:

- **The budget for it:** Most scalers cost over $1,000 (that's on top of your HDTV and source devices).

- **Multiple video sources:** If you're connecting just a new Blu-ray disc player to your fancy new 1080p HDTV, you might not have a strong use for a scaler. But most folks have regular DVDs, cable, satellite, or over-the-air standard- *and* high-definition TVs, VHS or D-VHS VCRs, DVRs, PCs, and even Internet-based content feeding into their HDTV. And all of these sources offer video of differing resolutions and qualities.
- **An internal scaler inside your HDTV that's less than perfect:** Manufacturers usually put pretty decent scalers inside of their HDTVs. But HDTVs are built to cost constraints, and they're also designed to be pretty thin and sleek. The scaler designs within them aren't no-holds-barred, top-of-the-line efforts in many cases.
- **A desire for the absolutely best picture you can get:** If you're looking for the ultimate picture, you might want to put the state-of-the-art in video engineering to work for you in this pursuit.

If you meet these criteria, well, you're in the target audience for a video scaler, so read on! And even if you're not, you might want to read on anyway for two reasons:

- These concepts apply to the internal scalers inside HDTVs, so they can help you understand that element of an HDTV as you're shopping.
- These scalers are nothing more than very powerful, sophisticated, and specialized computers. Like all computers, they get cheaper and better over time, so you might find that a scaler fits your budget and needs sometime in the future.

To make sure that a scaler can do its best for your HDTV, you need to do a bit of research into your HDTV's capabilities and features. Some HDTVs send every incoming video signal through its internal scaler — in effect, rescaling the video you've already scaled once in your external scaler. This second scaling process can cause picture degradation and reduce the effectiveness of your expensive scaler. Scalers work their magic best on HDTVs that allow *native resolution* or *1:1 pixel mapping* or *pass through* — three common terms that essentially mean that the display takes inputs that match up to its native resolution and leaves them alone, bypassing the internal scaler. If your display doesn't have this feature, you might still benefit from an external scaler, but you can get the best results if the HDTV itself takes the output of your scaler and displays it directly on the screen without further processing.

To complicate matters on this 1:1 pixel-mapping issue, some displays provide this feature only on certain inputs. For example, they might accept a native-resolution signal (like 1,366 x 768 for most plasmas) on the HDMI input but not on the component-video input. The thing to look for here is support for 1:1 pixel mapping on the input that you'll be connecting your scaler to. Remember that this isn't *mandatory* and you can still get benefits from external scalers, but external scalers work *better* when you can map the pixels coming out of the scaler directly to the pixels on your screen.

Fitting a video scaler into your system

If you're considering a video scaler, you might be wondering how to fit it into your system. After all, a scaler is neither fish nor fowl — it's sort of a unique beast that has a unique role and place in an HDTV home theater. Video scalers have four major functions, and understanding these functions gives you a mental picture of where they fit into the system:

- **De-interlacing video content for display on a progressive-scan HDTV:** All scalers de-interlace video sources — more importantly, they do it properly. Some HDTV internal scalers do a poor job of this task. (It's a lot of computation to take, for example, a 1080i video signal and de-interlace it.)
- **Scaling video resolution from the source or input resolution to the optimal resolution for the HDTV display:** Along with de-interlacing, resolution scaling is the main function of a scaler. (Given the name, this is sort of obvious, we suppose.) Most scalers scale all standard input resolutions (480i, 480p, 720p, and 1080i) to standard HDTV resolutions (720p and/or 1080i). A few scale to 1080p for HDTVs that accept this resolution. Some even scale to custom resolutions, such as 1,388 x 766 — the resolution of most 50- or 55-inch plasma HDTVs.

Why scale incoming HDTV signals (like 720p or 1080i) to other HDTV resolutions? The answer is simple if you recall that most HDTVs have an optimal native display resolution. So it might be better for you (in terms of picture quality) to take an incoming 1080i signal and convert it to 720p in your scaler if your HDTV's native resolution is 720p.

- **Further video processing of the signal to optimize it for display on the HDTV:** For example, scalers typically recognize when a video source is film-based, and they apply 3:2 pulldown correction to make this material look more natural. Many scalers can also do other processing tasks like handling 2:2 cadences (required for content that was created in the European PAL system), and also provide additional enhancements to the video signal's colors, correct noise and errors introduced by video-compression systems like MPEG-2, and help provide smooth pictures for moving images on-screen.

3:2 pulldown correction is just one of many cadences (think dancing, and the timing of your steps) that video processors need to deal with. They relate the conversions that video producers must make when they move content from, for example, 24 frames-per-second film to 30 or 60 frames-per-second video. To turn 24 into 30 or 60, some frames of video are repeated more often than others. A good video processor recognizes this situation and corrects any picture quality problems caused by these conversions (typically seen as unnatural moving images and blocky edges of objects on-screen). Although 3:2 pulldown correction is the most common cadence issue for scalers to deal with, it's not the only; video and film recorded in different parts of the world have their own unique cadences that have been converted to 30 or 60 frame-per-second video and therefore require their own unique cadence correction — and the best scalers can handle this with ease!

- **Switching between video sources to select the active source device and route its video to your display:** Most scalers can handle this routing for *all* video content, even content that doesn't need to be de-interlaced or scaled for your HDTV.

This last function provides the answer to the "where do I put my scaler?" question. A scaler should be installed *between* all of your video sources and your HDTV display itself. The scaler then processes the video (de-interlaces, scales, and applies other processing) and switches among sources to send the signal you want to your HDTV over a single cable. Most external scalers have a feature that automatically detects which source is actively sending a video signal and then routes that signal to your HDTV. You can also use your remote control to manually tell the scaler which source to route to your HDTV.

Looking more closely at scalers

Several brands of scalers are on the market. They're all aimed at the high-end HDTV and home-theater crowd, with prices to match. The good news is that prices are dropping, and at the same time, scalers are getting better and more powerful.

The earliest scalers didn't actually do any scaling in terms of converting one video resolution to another (like 480i to 1080p). Instead, they were simpler (though still quite complex for the day) *line doublers* that took interlaced 480i video content and translated it to a progressive-scan 480p output by taking each of the interlaced video fields, combining them, and then sending complete video *frames* to the display. In other words, they de-interlaced video. Line doublers predate HDTV and were most frequently used for expensive CRT-based front-projection TV sets used in home theaters back before digital displays became common and mainstream.

As line doublers became more powerful, they began to include additional *interpolation* capabilities that allowed them to not only combine odd and even interlaced video fields, but also use some computational *algorithms,* or mathematical models, to make up what might appear if more lines of resolution were in the video source. These devices, called line *quadruplers,* turned 480i video into 960p video; like line doublers, they were primarily designed for use in high-end CRT projection systems that could handle this resolution.

The advent of digital displays, with their fixed output resolutions, brought about a need for true image scalers that could convert various input resolutions into one display resolution. Unlike doublers or quadruplers, which would work only in multiples of 480 (or 576 for folks with PAL systems outside of the United States), image scalers use faster processing systems and improved algorithms. As a result, they can scale all sorts of input resolutions to specific display resolutions like 720p (1,280 x 720 pixels) or 1080i/p (1,920 x 1,080 pixels) and even to custom resolutions (like the 1,366 x 768 pixels that most plasmas have as a display resolution).

Despite these additional capabilities, image scalers are cheaper than line doublers and quadruplers used to be. High-end image scalers can cost less than $5,000, whereas image quadruplers often cost $20,000 or more in the late 1990s and early 2000s.

These are a few of the more popular image scalers:

- **DVDO:** The folks at DVDO (`www.dvdo.com`) make some of the most popular and well-received image scalers on the market. DVDO has three current models in its iScan series: the iScan VP20, VP30, and VP50 ($1,699, $1999, and $2,999, respectively). Each of these models can accept and scale 480i and p, 576i and p (PAL), 720p and 1080i signals over composite, component, S-video, and HDMI connections, and send the resulting scaled video signals to your HDTV by using component-video or HDMI connections. The fancier models (the VP30 and VP50) add some features, including de-interlacing of 1080i signals to 1080p, a greater number of input and output connections, and the ability to create custom output resolutions tailored to your display's requirements. The flagship VP50 model includes DVDO's well-regarded VRS Precision Deinterlacing Module, which provides motion adaptive de-interlacing that creates a much more realistic and film-like progressive-scan output.
- **Lumagen:** Lumagen (`www.lumagen.com`) has several models of image scalers ranging in price from $1,099 to $2,499. One of the more popular models is the VisionHDP, which sells for $1,499. This model accepts two DVI sources (these can be HDMI outputs from your source devices using an inexpensive adapter — just remember that DVI doesn't carry the audio signal, so you need to use separate audio cables), two component video inputs, two composite-video inputs, and two S-video inputs. It can scale and de-interlace input signals in 480 and 576i and p, 720p and 1080i as appropriate. It can configure the output to your HDTV up to and including 1080p (if your HDTV accepts this input). Many reviewers feel that the VisionHDP is a great bargain with a particular strength at processing film-based material (in other words, performing 3:2 pulldown correction).

- **Key Digital:** Key Digital (www.keydigital.com) offers several scalers, including the popular iSync HD, which sells for $1,750. This scaler accepts component, composite, S-video, and HDMI inputs, and it can scale 480i/p, 576i/p, 720p, and 1080i inputs and can output to all standard HDTV resolutions as well as custom resolutions for your particular HDTV (like 1,366 x 768 for most plasma HDTVs). An interesting feature of the iSync HD is its audio support — the iSync HD can output both stereo (left and right) analog and digital surround-sound signals to your receiver or audio processor, and you can use it as an audio-switching device.

Looking for HDTV Gadgets

Given the pace of change in the industry, here are the best places to check out any new HDTV gadgets that artificially stimulate your fancy:

- www.digitaldummies.com: Our companion site for all our books. Our site offers information about updates in HDTV topics, including new gadgets you should consider.
- Geeky gadget sites with the latest innovations:
    - www.gizmodo.com: This is Pat's favorite site.
    - www.engadget.com: If this looks similar to Gizmodo, that's because the site's editor used to be editorial director at Gizmodo.
- www.ehomeupgrade.com: This site tracks various developments in digital media gear, including streaming video and digital media servers — two areas where we expect a lot of change in HDTV.

Chapter 26 covers other accessories that you can tie into your HDTV system.

Part V
Sensory Overload

"HD TV, huh? Well'p there goes the ambiance."

In this part . . .

There's a lot to a good HDTV experience besides just a great TV set. Now, don't get us wrong, you want to start with a great TV set. But other components also drive your HDTV atmosphere, and we talk about all of these in this Part.

First, we delve into the whole audio experience, both with the built-in speakers in your HDTV and with a home-theater surround-audio system. We argue vehemently that *the whole point* of going HDTV is to create a total immersion in the content, and you can't do that without real surround sound.

Then we talk about your HDTV room — nay, your HDTV theater — and how you can optimize that for the video and audio that comes with high-definition content. We advise you on placement, construction, wall coverings, lighting, and the like.

Finally, we talk about how to enhance and optimize your HDTV experience. Keep in mind that you can't judge an HDTV by its display. There's so much about an HDTV system that is more than skin deep, and this part of the book helps you understand how to optimally install, calibrate, and scale your HDTV system.

By the time you're done with this part, you can rest assured that your HDTV is just the way it should be. Isn't that a comforting feeling!

Chapter 17

Understanding Audio

In This Chapter

- Figuring out what's watt in audio
- Surrounding yourself with sound
- Deciphering HDTV audio specs
- Surrounding yourself with only two speakers

Of course, the picture — especially the big-screen picture — provided by HDTV content shown on HDTV-capable displays is the number-one attraction. But — and we think this is a *big* but — high-quality audio reproduction comes in a close second when it comes to making HDTV more of a "you're-there" experience. In other words, good audio (particularly, good *surround-sound* audio) lets you become more immersed in the HDTV experience. The audio helps make your HDTV viewing seem more like that proverbial window to the world.

In this chapter, we give you good, quick info on the confusing specifications and terms that you deal with when trying to understand the capabilities of an audio system. Some audio manufacturers fling bovine byproducts when they describe (dare we say *overstate?*) their audio capabilities, so you need some knowledge about what's actually what in your audio systems.

We help you dig through the maze of surround-sound standards to understand which speakers perform what function. We also cover the built-in audio systems found on most HDTVs.

Chapter 18 covers external surround-sound systems. They're the *best option* for a great HDTV system. Most built-in HDTV sound systems are okay for casual viewing, but not so good when you really want to get into a movie.

Grasping Audio Basics

Before we get into any specifics about particular systems or audio components, indulge us by reading a few paragraphs about audio specifications.

Yes, we know, reading the specs themselves is bad enough — now we want you to read *about* reading the specs? Yikes! Trust us, though; we have a method underlying our madness.

Audio specs are some of the most misused numbers in the world. There isn't really any enforced standardization for how manufacturers measure and report the numbers behind their audio. We're talking mainly about audio power specs (watts).

For example, two manufacturers can both claim that their systems put out 50 watts per channel (a decent amount). Neither manufacturer is *lying* when it states this specification — but one system might be much more powerful than the other.

How is this possible? Well, the simple answer is that people have different ways of measuring the same thing (watts, in this case). It's sort of like measuring feet (the distance) with your feet (the appendages) — everyone's feet are different sizes, so *x* number of Pat feet doesn't equal *x* number of Danny feet. So if two manufacturers are using different-sized "feet" to measure their audio systems, Brand X's watts might not equal Brand Y's.

Here's an explanation of audio measurements:

- **Watts:** The most basic measurement of an audio system — the number that gives you some idea of how loud the system is — is the power rating in watts. All else being equal, a system with a higher wattage rating plays louder. Wattage is measured in watts *per channel* (or speaker).

  Keep in mind that it takes a large increase in watts to make a truly noticeable difference in volume. To make a system play *twice* as loud, you have to increase the wattage by approximately *four* or more times.

- **THD:** Audio-system wattage is measured at a certain level of *distortion* (noise introduced by the audio amplifier system) called *THD* (total harmonic distortion). As audio systems are pushed closer to their volume limits, they tend to produce greater amounts of distortion. Manufacturers can make a system seem more powerful by measuring watts at a higher THD. Look for receivers that meet your wattage requirement when measured at low THDs, like 0.02 percent, rather than higher ones like 0.2 percent or even 1.0 percent.

- **Full-bandwidth power ratings:** Another gray area in power ratings is the frequency range at which watts are measured. The human ear can hear audio signals between 20 and 20,000 Hz. It's best if the system's power is measured across this entire range. Some manufacturers provide wattage ratings at only one frequency (such as 1,000 Hz), which can create an artificially high power rating. Try to find specifications that cover the full 20-to-20,000-Hz range to make true comparisons.

- **Ohms:** The *impedance* (or resistance) of the speaker being driven also affects the number of watts a system can produce. Manufacturers generally measure impedance at 8 Ohms, but sometimes you see wattage measured at lower impedances, such as 4 or 2 Ohms. It's good that a receiver can drive speakers with such low impedances (not all can), but the wattage measured at 4 Ohms is higher, so it shouldn't be compared directly to wattage measured at 8 Ohms.
- **Power handling:** The power-handling rating (also measured in watts) relates to the speakers in an audio system, not the amplifiers. This is simply a measure of how many watts the speakers can take before they start to shred themselves into confetti.

  Power handling is not — we repeat, not — in *any* way a measure of how loud the system is. The main speaker measurement regarding loudness is the *sensitivity* of the speakers — a measure of how much volume the speakers put out with a certain wattage of input from the amplifier.

The bottom line here is to make sure that you're comparing apples to apples as you look at audio systems — either built into an HDTV or in a separate home-theater receiver system. To go back to our earlier example, two manufacturers can have 50-watt systems, but one might be measured at a limited bandwidth, at a high THD, and on a lower-Ohm impedance — and might be significantly less powerful than the other.

Surround-Sound Mania

The profusion of *surround-sound* standards is a confusing area in the audio arena. Surround sound is multichannel audio designed to produce *spatial audio cues* — sounds from all around you, in other words — relating to action on the HDTV's screen. It's usually described by the number of channels (or speakers) that a particular system uses to envelop you in sound:

- 5.1 channel system audio actually has six channels:
  - A center-channel speaker (located directly above/below your HDTV) that reproduces dialogue on your screen.
  - Two front (or main-channel) speakers, which reproduce most of the musical soundtrack, plus left and right spatial cues (like someone walking into the room from one side or the other).
  - Two surround speakers, located on the rear side walls of the room, that produce spatial cues *behind* you, and also provide *diffuse* (not easily locatable) sounds to help create an audio atmosphere.

- An *LFE* (or low-frequency effects) channel that uses your system's subwoofer (if you have one) to reproduce the very deep bass notes and sounds (like cannons exploding).

  The LFE channel, because it contains only a small portion of the full spectrum of audio frequencies humans can hear, is the ".1" of 5.1 (or any *x*.1 system).

- 6.1 channel systems add one extra speaker — a *rear surround* that is usually located on the back wall of your HDTV viewing room. It provides an extra level of surround-sound detail.
- 7.1 channel systems add two extra speakers, mounted on the back wall of the room.

Figure 17-1 shows a 5.1 channel surround-sound layout in a typical HDTV viewing room or home theater.

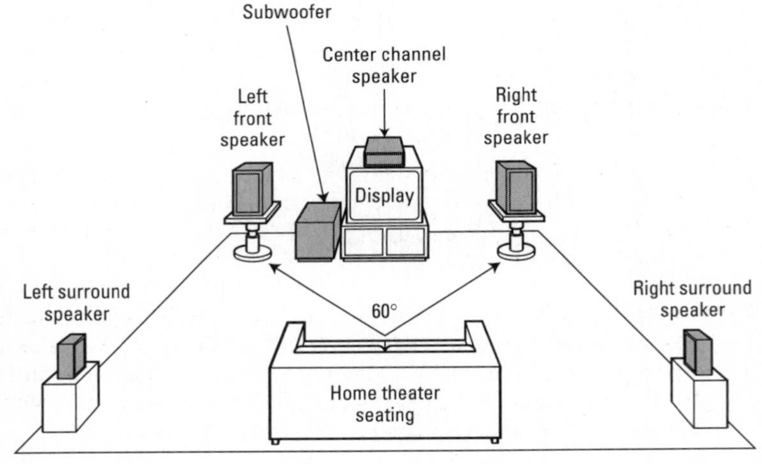

Figure 17-1: Surround yourself with sound.

Understanding how surround sound is created

A surround-sound audio signal can be created three ways:

- **It can be encoded in a DVD or HDTV program *discretely.*** With this method, each individual channel is recorded on its own channel within the audio soundtrack. This is the best way to accurately produce surround sound — when the director wants you to hear that spooky footstep *right there,* discrete surround sound gives you the best chance of hearing it there.

- **It can be *matrixed* in along with other audio channels.** Typically, matrixed surround-sound signals are mixed into normal two-channel *stereo* soundtracks. If you're listening in stereo, you don't even hear the surround-sound cues, but if you have a surround-sound system, these "hidden" tracks are extracted from the stereo soundtrack.
- **Sometimes there is no surround sound (discrete or matrixed) for an HDTV program.** Perhaps the movie was filmed in 1942, when there wasn't even stereo! Well, surround-sound hardware can often create its own best guesstimate of surround sound, using a regular two-channel stereo input.

If you have a piece of physical media, such as a DVD, you can usually figure out which surround-sound *format* (or system) is being used by looking for labels. (We discuss the different formats in the following section.) If you're watching a broadcast in HDTV or standard-definition, you might see a label or description on an on-screen program guide (or in the paper or *TV Guide*). Luckily, your surround-sound *decoder* — the device that reads the encoded surround-sound signals and turns them into sounds you can hear — usually detects surround-sound formats automatically.

Understanding the formats

Two companies dominate the surround-sound system market: Dolby Labs (`www.dolby.com`) and DTS (`www.dts.com`). Dolby is the market leader, but both are common on DVDs and other source material. These are the most common surround formats:

- **Dolby Digital/AC-3:** The most common surround-sound format, Dolby Digital (also called AC-3) is part of the HDTV standard itself! Dolby Digital is a 5.1 channel, discrete surround-sound format, and in addition to HDTV programming, it's on most DVDs and on some digital cable and satellite TV programming.

  Dolby Digital doesn't *have* to be 5.1 channels. It's possible to use Dolby Digital for two-channel stereo or even mono (one-channel) soundtracks — which is often the case for older material filmed/recorded before the advent of surround sound.
- **Dolby Digital EX:** As Dolby's 6.1 channel solution (with the extra rear-surround channel added in), Dolby Digital EX provides 5.1 channels of discrete surround sound, but then uses a matrixed system for the rear surround.
- **Dolby Pro Logic II/IIx:** Pro Logic II is Dolby's system for decoding the matrixed surround sound found on some older TV sources — like VHS VCR tapes and some stereo NTSC TV broadcasts. A newer version named Pro Logic IIx is starting to appear on some audio/video gear. Pro Logic II/IIx can also create relatively realistic-sounding surround sound from true two-channel sources like CDs or stereo TV broadcasts.

- **DTS:** DTS (the format) is DTS's (the company's) equivalent to Dolby Digital — a 5.1 channel surround-sound format. You mainly find DTS on DVDs.
- **DTS-ES:** DTS-ES is DTS's equivalent to Dolby Digital EX; it's a 6.1 channel system. Found on DVDs, DTS-ES differs from Dolby Digital EX in that at least *some* DTS-ES soundtracks use a discrete rear-surround channel. Not all do, however — look for the DTS-ES Discrete logo on the DVD case; otherwise, assume you have a matrixed DTS-ES soundtrack.
- **DTS NEO:6:** Not content to have equivalents to only Dolby Digital and Dolby Digital EX, DTS also has a system equivalent to Dolby Pro Logic II: DTS NEO:6. It takes two-channel audio input and magically creates multichannel 5.1 (or even 6.1 or 7.1) surround-sound soundtracks.
- **Proprietary encoders:** Some HDTVs (and other surround-sound gear) have a surround-sound system from someone other than Dolby or DTS. These systems typically provide functionality similar to Pro Logic II or NEO:6, creating multichannel audio output from two-channel input.

So why should you care? Well, you have two issues here:

- Be sure your system can decode the right audio formats.
- Buy the versions of audio that are most compatible with your system (or else know what you *aren't* getting when you buy).

In an ideal world, your audio-playback capability would support all these formats, but you want at least Dolby Digital. Most systems support both Dolby Digital and DTS. As far as the 6.1 or 7.1 channel systems go, that's a matter of personal taste. Not a ton of content is out there for them, so they aren't requirements.

You can tell what encoding was used to create the audio by looking on the back of the DVD or CD case.

Introducing the new contenders

Surround sound doesn't end with Dolby Digital and DTS. The launch of the new high-definition optical disc formats (HD DVD and Blu-ray, discussed in detail in Chapter 11) has brought three new and improved surround-sound formats that go beyond Dolby Digital and DTS in two major ways:

- **They offer more channels of audio for a more immersive surround-sound experience.** Depending upon which system is used, and how it is implemented, you can get up to 14 channels (13.1, in other words) of audio. Your biggest problem won't be getting great surround sound — it'll be finding a spot for all the speakers.

The 13.1 channel sound is a part of the specifications for these new surround-sound formats, but the systems that are shipping today and in the near future are limited to 7.1 channels. The extra channels are built into the formats for future use. But the good news is that all 7.1 of these channels can be discrete (today's 7.1 channel systems use matrixed sound for some of the channels), so you get a more accurate reproduction of the surround-sound intents of the creator of the programs you're watching. So today you get better surround sound, and some time in the future, you'll get more of it!

- **They offer higher-quality sound.** This goes beyond the discrete (non-matrixed) sound. These new formats also are designed to provide more *bits* to their digital audio signals. That means that these formats use *less* (and in some cases, effectively *no*) compression that can cause the sound quality to be degraded. If you're familiar with the MP3/iTunes/Windows Media world of digital audio, this concept might be familiar to you. You can compress a song down to a very low bit rate to save space on your disc, but the sound quality suffers. Similarly, older surround-sound formats compressed the audio to fit within the confines of a DVD, but the new ones take advantage of the additional space on HD DVD and Blu-ray discs and use less compression to give you better sound.

We're stretching the truth a bit by saying that some of the new formats use effectively no compression of the audio signals in a surround-sound soundtrack. In fact, these formats all use compression, but two of them use a type of compression known as *lossless* compression. Lossless compression systems reduce the amount of data needed to store audio signals, but do so in a way that keeps all of the original data intact. When you play the audio, it's an exact duplicate of the original recording. On the other hand, *lossy* compression systems (which are used in all other surround-sound formats) throw away some of the data representing the original audio signal in order to save even more space on the disc. Sophisticated *algorithms* (computational processes) within these lossy compression schemes reduce the audible impact of this discarded audio data. However, for the utmost in audio fidelity, lossless is the way to go.

You should be familiar with these three new surround-sound formats:

- **Dolby Digital Plus:** A big upgrade to the older Dolby Digital surround-sound formats, Dolby Digital Plus uses lossy compression, but at much higher bit rates (up to 6,144 Kbps, versus Dolby Digital's 640 Kbps), so a lot less compression is required. Dolby Digital Plus also supports up to 13.1 channel audio (though the first versions on the market are limited to 7.1).

  Dolby Digital Plus is a mandatory format for HD DVD and optional for Blu-ray players (so many players, but not all, support this format), and it might eventually be found on broadcast HDTV programming as well.

- **Dolby TrueHD:** The *crème de la crème* of Dolby's surround-sound formats, Dolby TrueHD provides lossless audio for up to 13.1 channels of surround sound. Wow! When we heard Dolby's demos of TrueHD, we were just blown away. Like Dolby Digital Plus, TrueHD is mandatory for HD DVD hardware and optional for Blu-ray. We don't expect to ever see it used for broadcast HDTV, simply because of its bandwidth requirements.
- **DTS-HD Master Audio:** The folks at DTS have their own new surround-sound format in DTS-HD Master Audio. Like Dolby TrueHD, this is a lossless format that supports 7.1 or more channels of super-high-quality audio. DTS-HD Master Audio support isn't mandatory in either HD DVD or Blu-ray, but it's optional in both devices.

To take advantage of these newfangled surround-sound systems, you need a few items:

- An HD DVD or Blu-ray player that supports these formats
- An HD DVD or Blu-ray disc formatted in one of these formats
- A surround-sound audio system (these are discussed in Chapter 18) that supports the appropriate format

The last piece of this puzzle hasn't, as we write in mid-2006, hit the market yet. Eventually (or, most likely, by the time you read this), new home-theater receivers with HDMI inputs (we talk about HDMI in Chapter 3) will be ready to decode these new formats.

Until those receivers hit the market, you have two alternatives:

- You can let the decoder inside your HD DVD or Blu-ray player handle the surround-sound work and then send multiple analog audio signals to your receiver.
- You can rely upon Dolby Digital Plus and DTS-HD Master Audio's *backwards compatibility.* These formats both work with older home-theater gear that supports Dolby Digital and DTS, respectively. You don't get the full quality possible, but you can get a great surround-sound experience that's better than what was possible with previous formats.

Dealing with Built-In Audio

Most — but not all! — HDTVs have a built-in audio system. TV audio systems have never been known for their audio fidelity, but you'd be amazed at how good some of these built-in audio speakers sound.

Even if you have a fancy home-theater audio system, you might find that the built-in audio system on your TV is good enough for casual viewing — like watching the evening news or listening to the *Today Show* while cooking breakfast in the kitchen. In fact, at times like this, it's probably preferable to just fire up the HDTV and use its built-in speakers, rather than warming up the mondo complicated home-theater system. When using the TV with a surround-sound system, the TV's speakers usually assume the role of a center front speaker.

Here's what we think you should look for when you evaluate the built-in audio in an HDTV system:

- **Amp power:** Although amplifier power ratings for TVs are sometimes all but useless, you should look for a system that puts out 10 to 20 watts per channel, if you're going to rely only on the TV's audio system.

  Check out the THD (see the earlier section, "Grasping Audio Basics" if you aren't familiar with THD) and other power-rating factors when you examine this number. TV wattage ratings are often measured at significantly higher THD levels than "real" audio equipment is. So that "50-watt" TV system might be equivalent to a 10-watt home-theater receiver, powerwise.

  Some manufacturers claim *x* watts, but they aren't talking about individual channels. Instead, they're referring to the sum total of all amplifier totals. You certainly can't compare this number to the watts-per-channel rating of an external audio system.

- **Number of speakers:** Most HDTVs include two speakers: left and right. A few have some additional speakers that you can place throughout the room, but this feature is increasingly rare, particularly in light of the low prices of home-theater systems these days.

- **External speaker attachments:** Some HDTVs don't have extra speakers but have amplifiers that can power external speakers. In this case, you attach your own speakers to the back of your TV with some standard speaker cables. (This is another rare option.)

- **Built-in surround-sound decoder:** True HDTVs — those with built-in HDTV tuners — include a surround-sound decoder that can decode the Dolby Digital signal used by HDTV broadcasts.

  Most HDTVs only have two speakers, so they won't create surround sound from an HDTV broadcast. The decoder is just there so the HDTV can turn the surround sound into two-channel audio, or feed surround channels to an external system.

- **Special two-channel surround modes:** Many HDTVs include special audio circuits that can help create the *illusion* of surround sound from the two speakers built into the TV. (See the sidebar titled "Creating surround sound from thin air!")
- **Connectivity to home-theater receivers:** If (*when,* in our opinions) you decide to move beyond the two speakers built into your HDTV, you'll want to connect your TV to a home-theater receiver. If you have an HDTV with a built-in HDTV tuner, it needs a *digital audio output* to connect to your receiver. (Chapter 3 covers digital audio outputs.)

  This connection is almost standard equipment for any true HDTV, but you should make sure. Look for either Toslink *(optical)* output or coaxial digital output.

Creating surround sound from thin air!

You need at least 5 (.1!) speakers for real surround sound in a viewing room, but your brain can *think* you're surrounded by speakers when you aren't.

For example, the folks at Dolby Labs have a couple of systems (which you might find in some HDTVs' audio systems) that make two speakers sound like 5.1 or more! The two systems are *Dolby Headphone,* which reproduces surround sound for headphones, and *Dolby Virtual Speaker,* which does the same with two conventional speakers. These systems use computer horsepower to modify the sound going to your two speakers (or headphone transducers) by adding *echoes* and *delays.* These echoes and delays are designed to reach your listening position so your brain is fooled into thinking that it hears more than two speakers.

Are these systems any good? Well, they sure aren't bad. Dolby Virtual Speaker and SRS Labs' *TruSurround* can do a good job of fooling you. If you aren't installing a real surround-sound system, look for an HDTV with a system like these.

We don't think you can substitute these virtual surround-sound systems for even a small, inexpensive, real surround-sound system (such as a $200 Home Theater in a Box system). But if you don't have the room or budget for 5.1 speakers, give virtual surround sound a whirl and see what you think.

Chapter 18

Home-Theater Audio

In This Chapter

- Moving your audio out of the TV
- Checking out all-in-one systems
- Getting into home-theater receivers
- Choosing features

In the preceding chapter (Chapter 17), we talk about surround-sound audio. Surround sound is an integral — and, we think, essential — part of HDTV. And the best way to get *real* surround sound is to leave behind the built-in HDTV sound system (the speakers that came installed in your HDTV, in other words) and connect a home-theater audio system to your HDTV.

In this chapter, we discuss some options for adding an external surround-sound system to your HDTV theater. First, we show you the popular Home Theater in a Box solutions that include the speakers, amplification, surround-sound decoding, and often a DVD player all in one box. These solutions are the easiest way to add surround sound to your HDTV.

We also talk about build-it-yourself systems with separate speakers and parts from different manufacturers theater receivers. These systems take a bit more work on your part to assemble, but they potentially offer better sound because you can mix and match the best pieces.

Boxing Up Your Home Theater

Home Theater in a Box (HTIB) systems are the simplest, quickest, and easiest way of adding surround sound to your HDTV. With HTIB, you go to the store (or shop online) and bring home a single box. Everything you need — usually including all the cables and wires that connect everything — is right in that box, ready to go.

HTIB systems are prematched, preconfigured, and prepared for quick and easy installation. The price you pay for this convenience is a lack of flexibility — you don't get to pick and choose individual components on their merits. HTIB systems can sound very good but fall a bit short of slightly more expensive systems where you choose your own components (such as speakers, receivers, and DVD players).

Typically, an HTIB contains the following:

- **Amplifiers for at least five channels:** The amplifiers take the low-level audio signals recorded in an HDTV program or on a DVD soundtrack and amplify them electrically so that they drive your speakers. Look for at least 50 watts per channel.
- **Surround-sound decoders:** The surround-sound decoder extracts multiple audio channels from the soundtrack of an HDTV broadcast or DVD (or any audio source) and directs the audio signals to the appropriate amplifier channels.
- **An AM/FM tuner:** For tuning in radio broadcasts.

  Together, these first three items make up a home-theater receiver (discussed in the next section, "Receiving Home-Theater Sounds").
- **A DVD/CD player:** Most HTIBs include a DVD player, either in the same chassis as the amplifier and surround-sound decoders, or in a separate chassis that connects via standard audio and video cables. Turn to Chapter 11 to find out more about DVD player features and specifications.

  As we write, *no* HTIB systems include a Blu-ray or HD DVD player for getting high-definition movies into your HDTV. This isn't surprising when you consider that only a handful of these players are on the market in any form. We expect that in a year or two, high-definition HTIB systems will become available; for now, you need to go with separate components to take advantage of these latest disc formats. We discuss separate components in the next section.
- **Five or more speakers:** HTIB systems typically include five small *satellite* speakers. (They're called satellites because they hang around your subwoofer like Sputnik orbiting Earth.) For 6.1 or 7.1 channel HTIB systems, expect an extra one or two speakers.
- **A subwoofer:** This speaker — usually the largest in your HTIB system — handles the lowest of the low-end frequencies in your surround-sound system.
- **Wires and cables:** Here's where HTIB makes things really easy — most systems include all the cables you need, and they're usually very clearly labeled, so it takes absolutely no brain power to hook things up.

In the section that follows, we discuss a variety of features to consider when choosing a home-theater receiver. Most of these features also apply when choosing an HTIB system.

Receiving Home-Theater Sounds

HTIB systems can be great, and they sure are convenient, but the fact is, the companies that are best at making receivers usually are *not* also the best at creating speaker systems. Sure, those really huge consumer electronics manufacturers have a lot of talented engineers on staff, but we tend to prefer speakers from specialist companies who focus solely on speaker design and manufacture. Going with your own components lets you do this mix-and-match thing.

The centerpiece of a surround-sound audio system is the *home-theater receiver* (also called an *A/V receiver* or just a *surround-sound receiver*). The receiver's many duties are the following:

- Decoding surround-sound signals from DVDs and other sources
- Amplifying audio signals and sending them to the speakers
- Switching between different audio and video sources
- Adjusting audio volume and tones
- Tuning in radio broadcasts
- Acting as the main user interface to the home theater (which is its most important function)

Literally hundreds of home-theater receivers are on the market these days — ranging from under $200 to well over $4,000. When we're shopping for a receiver, here are the items we make sure to check off on our list:

- **Power:** We like to look for receivers with at least 70 watts per channel, but we prefer 90 or 100 if the HDTV is in a larger room.
- **Number of channels:** All home-theater receivers can support five channels of audio. Some support six or seven channels for a 6.1 or 7.1 system. (See Chapter 17 for details on surround-sound channels.)
- **Surround-sound modes:** A baseline system supports Dolby Digital and DTS. Many receivers also support 6.1/7.1 systems such as DTS-ES. We also like to look for Dolby Pro Logic IIx support in a receiver, for providing simulated surround sound for mono and stereo content (such as older TV shows).

The new Dolby and DTS formats (Dolby Digital Plus and TrueHD, and DTS HD) aren't currently supported in any home-theater receivers. To use these formats (which can be found on HD DVD and Blu-ray discs), you need to use the surround-sound decoders built into the HD DVD or Blu-ray player, and you need to connect to your receiver with analog audio connections (see Chapter 3). Eventually, receivers with HDMI connections will be equipped with surround-sound decoders for these new formats.

- **On-screen display:** Most receivers have an on-screen display that lets you see and control your receiver's settings by accessing a user interface that's transmitted to your HDTV's screen. This is a whole lot easier than trying to see the controls on the tiny display on the receiver itself.

  Make sure that your receiver's on-screen display works over component video or HDMI connections. Some receivers send the on-screen display only over composite or S-video cables. That means you can't access the on-screen display while you're watching high-definition sources — which kind of defeats the purpose for us!

- **Automated surround-sound equalization:** Many surround-sound receivers that include automated systems use a supplied microphone to automatically set up the proper audio levels for your room. (We talk about this process in Chapter 19.) With these automated systems, you can put the microphone where you normally sit, and a computer in the receiver goes through and automatically adjusts your receiver's settings to customize the audio for your room. It's very handy and a big time-saver.
- **Video switching:** Video switching allows the receiver to act as the traffic cop for your HDTV system — taking all video source signals and routing them appropriately. Having a receiver that can do video switching is a big plus if you have more components to connect to your HDTV than you have connections available on the TV itself.
- **Component-video support:** As we discuss in Chapter 3, HDTV signals require at least component-video connections, but not all home-theater receivers support component-video switching. Look for a receiver with component-video *bandwidth* of at least 30 MHz if you want to run HDTV through the receiver.
- **Video up-conversion:** You probably have a variety of different video sources feeding into your HDTV through your receiver. Some use component video, others S-video, and still others composite video (such as a VHS VCR). Video up-conversion takes all those sources and converts them to be carried over a single cable or set of cables to your TV. For example, component-video up-conversion allows you to send all your video to the HDTV by using a single set of component-video cables. Such a setup makes it easier to connect your HDTV to your receiver (you need fewer cables), and makes it easier to set up your remote control and HDTV because all video comes in over a single set of cables.
- **HDMI support:** A growing number of home-theater receivers support HDMI switching. (We talk about HDMI in Chapter 3.) HDMI switching lets you take a number of HDMI source devices in your home theater (such as a high-definition DVD player, set-top box, or game console) and connect them through the receiver to your HDTV. The great thing about HDMI as a connection method is that it carries both high-quality digital video *and* digital surround-sound audio. So you need only a single cable

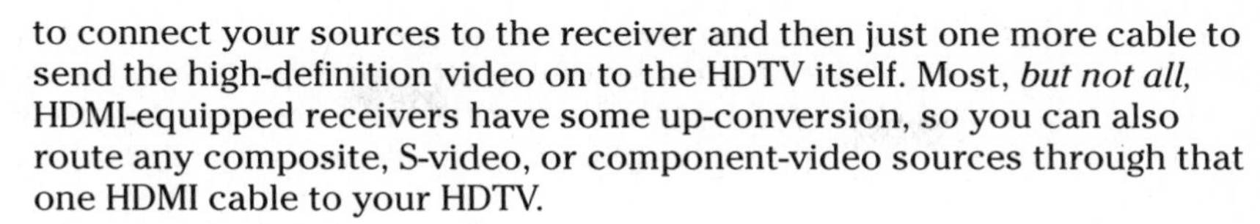

to connect your sources to the receiver and then just one more cable to send the high-definition video on to the HDTV itself. Most, *but not all,* HDMI-equipped receivers have some up-conversion, so you can also route any composite, S-video, or component-video sources through that one HDMI cable to your HDTV.

If you're buying a new receiver, we highly suggest that you go for one with HDMI switching and — if your budget can swing it — HDMI up-conversion.

Some receivers have HDMI switching and component-video up-conversion that up-converts all of your analog sources but uses a component-video connection for this up-converted signal. In this case, you need two sets of cables between your receiver and TV — one HDMI cable for HDMI sources, and one set of three component-video cables for all your other source devices.

- **Number of inputs:** Take a look at the back of your home-theater receiver (see Figure 18-1 for a typical view) and count up the number of inputs (both audio and video; we describe both types in Chapter 3). The key here is to make sure that your receiver has enough of these inputs to accommodate all the gear you might want to connect to your HDTV system.

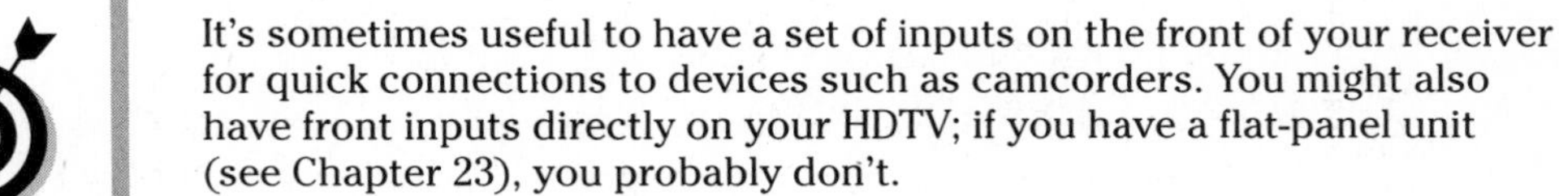

  It's sometimes useful to have a set of inputs on the front of your receiver for quick connections to devices such as camcorders. You might also have front inputs directly on your HDTV; if you have a flat-panel unit (see Chapter 23), you probably don't.

- **Assignable inputs:** One really cool feature found on an increasing number of receivers is *assignable inputs.* Instead of rigidly matching the receiver's digital-audio inputs to a particular video input, you can use assignable inputs with any video input on your system. This just makes things a whole lot more flexible when you're trying to connect multiple devices that use digital audio (and there are many, such as HDTV tuners, DVD players, D-VHS VCRs, and even game consoles such as Xbox).

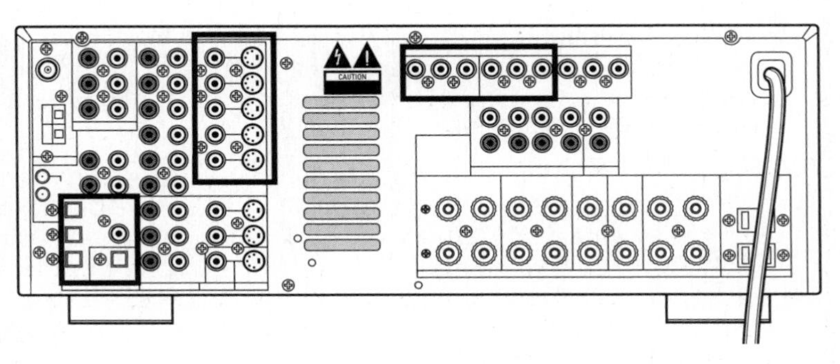

Figure 18-1: It's a jungle back here. Check out all the receiver inputs/ outputs.

The biggest disadvantage of skipping over the HTIB solutions (and going with a separate receiver/DVD/speaker systems) is that you have to spend a little bit more time shopping for gear. You have to pick out that receiver, a DVD player, and a set of speakers. We personally don't find this too much of a hassle. (Heck, we even wrote another book, *Home Theater For Dummies,* on just that subject!) If you do find it a hassle, check out the HTIB systems — they pack a surprising bang for the buck.

Going with separates

Moving from an HTIB solution to your own set of surround-sound speakers and home-theater receiver is a step up, providing (potentially at least) better sound quality. The next step is to move from a home-theater receiver to *separates.*

In a separates system, the functions of the home-theater receiver are divided up among — no surprise here — separate components. A separates system usually includes the following:

- **Power amplifiers:** These devices provide the actual audio signal amplification that "drives" the speakers in the surround-sound system. The power amps are the brawn of the system.
- **Surround-sound controller/preamp:** The controller/preamp handles all audio and video switching — sending audio/video inputs to the HDTV. The controller also decodes surround-sound formats and "steers" audio signals to the correct channel of the amplifier (and therefore to the correct speaker). The controller is the brains of the system.
- **AM/FM tuner:** If you want *all* the functions of a home-theater receiver, you have to add a separate AM/FM radio tuner. Nope, this has nothing to do with HDTV, but we just thought we'd be complete!

So why are separates an improvement over all-in-one receivers? Well, the difference is subtle, but separates are better at these tasks:

- Carrying the amplification and control/pre-amplification tasks in a separate chassis can be a benefit all in itself. Keeping these functions electrically separate — and using independent power supplies for each — reduces the risk of interference and noise getting into your surround sound. This is, however, a subtle difference.
- Separates allow you to choose the system that works best for you. Need more power for a large room? Get a bigger power amp. More concerned with a feature such as video up-conversion? Get a fancier preamplifier/processor.
- Separates make it easier for you to upgrade over time (for example, when a new surround-sound format hits the market). We know lots of people who are constantly upgrading their systems on eBay, selling their old stuff, and buying someone else's *newer* old stuff.

Chapter 19

Setting the Mood

In This Chapter

- Creating the right atmosphere through light and sound control
- Optimizing walls, floors, and ceilings for HDTV
- Automating your HDTV theater
- Putting your money where you park it

When you go to the movies at your local cinema, you can't help but notice the little things they do to try to create a certain atmosphere. There are dimmed ceiling lights, moving curtains, aisle side lighting accents, and nice seats — all the creature comforts you need to watch the creature feature.

However, a lot more is going on there than you might think. The large video screens, surround-sound audio, tiered chair placement . . . these are painstakingly planned to give you an uninterrupted view of the movie content. By minimizing things that would remind you that you're in a theater, movie-theater planners create a grand illusion that you are right up there on the screen with the characters, in the thick of the action.

Buying an HDTV *can* be simply about taking the TV out of the box, plugging it in, and then watching the great picture. But why stop there? Your HDTV experience could be a whole lot more exciting with the right equipment placement, lighting, sound treatments, and other nice touches that turn a living room or spare den into your true HDTV theater.

Choosing Your HDTV's Home

Turning your HDTV room into an HDTV theater is all about trying to optimize that same sense of illusion found in the movie theater. You can definitely put your HDTV in the wrong place, even if you had your heart set on a particular spot next to the ficus plant. Here's a list of things to think about as you locate your HDTV:

- **Room layout:** You tend to get more awkward sound patterns in perfectly square rooms. The best place to put your HDTV is along the short wall of a rectangular room, preferably a wall without windows or doors on it. Fully enclosed rooms are best for sound. If you must use a room that's open to another room, consider pulling heavy curtains across an open wall when you're watching films in your HDTV theater.
- **Seating layout:** Many people place a couch against a wall, with the HDTV in front of it. With a surround-sound-enabled HDTV theater, you want enough space behind you so the sound can get behind you and truly surround you. So the ideal position for your HDTV seating is a location more central to the room, or at least set off from the back wall.
- **Stray lighting and noise:** Stray light in a room, day or night, can substantially affect the HDTV viewing experience. The same is true for stray noise. Think about how lights from other rooms or street lighting might affect the ambience of the whole room. Listen closely to your room for regular interfering sounds, such as a clock ticking or a fish-tank pump whirring. Consider moving these devices if you can. And if the light or sound is coming externally (say, from a floodlight or a dryer or washer), consider some cheap absorptive wall coverings to shield and muffle them.
- **Distance to the picture:** There's a correct viewing distance and maximum viewing angle for HDTVs. (We explain viewing distance later in the sidebar titled "The right distance from the picture.")

  A TV can be too big or too small for your room, and you should sit within its best angle of viewing.
- **Reflected sound and light:** Think about how video and audio signals behave in your HDTV room. Muted wall colors or irregular wall coverings — bookcases are ideal — absorb stray light. A dark gray or black room is best, or one with heavy, colored drapes. (Now you know why you see all those drapes and carpeted walls in theaters!)

  Avoid mirrors, picture frames, and brightly colored gloss paint. They reflect light that creates *light ghosts* on the sides of the screen.

  Bare tile or wood on the floor causes acoustical reflections that mess up your *sound field* (the total sound "picture"). A good rug can absorb stray sounds that would otherwise muffle the crispness of your sound system.

  It's okay for the back wall to be a little reflective — it helps build a general sound field behind your seating area.

Creating the right atmosphere for your HDTV can be pretty expensive, but only if you want it to be so. In the rest of this chapter, we show you some simple lighting, sound, automation, and comfort concepts. Some of this is easier to do if you're building a room for your HDTV theater from scratch, versus merely sprucing up an existing room. In all instances, you can spend a wide spectrum of money, from the very inexpensive treatments to very involved construction. You can choose your pain.

The right distance from the picture

Obviously, where you sit to watch your HDTV is a very subjective thing, but it's a topic that some people have spent a lot of time thinking about. (What's the title on their business cards? "Deep TV-Distance Thinker?")

HDTV screens are very different from regular TVs in one big way — because they have such high resolution, it's harder to discern the visibility of the scan lines, so you can move in closer to the screens if you want to. Crutchfield (`www.crutchfield.com`), always a good source for audio and video information, recommends the following distances from your HDTV:

Viewing Distance	Recommended Display Size
5 feet, 7.5 inches	27 inches
6 feet, 9 inches	30 inches
8 feet	35 inches
9 feet	40 inches
9 feet, 9 inches	42 inches
10 feet	45 inches
10 feet, 5 inches	50 inches
12 feet, 6 inches	55 inches
13 feet, 9 inches	60 inches
15 feet	65 inches

We can't say it enough: If you really want to set up a room specifically for your HDTV— a home theater in your home — consider getting our *Home Theater For Dummies,* 2nd Edition, in which we tell you more about specific paints, room construction, and other tips for making it just right. Although there is some overlap with this book, *Home Theater For Dummies* is more focused on creating a specific environment for your HDTV to sit within.

Sounds Right?

Most people who shop for audio gear for their HDTV setup focus on how many watts this one puts out, how many interfaces that one has, and so on. What most people don't realize is that sound quality has at least as much to do with room acoustics as with the gear.

The room shape, flooring, wall coverings, and furniture all have a massive impact on the quality of sound from your HDTV theater. The sound is getting from point A (the speakers) to point Z (your ears) through a number of intermediate points — including some sound-reflecting surfaces throughout the room.

Because you're likely to put your HDTV in an existing space, you're probably starting at a decisive disadvantage when it comes to optimizing a room for sound. We often see an HDTV in a living room that opens into the kitchen or dining room in the rear — so no rear wall can reflect sound. This substantially affects the aural sound field in a surround-sound system.

We talk about sound a lot because we think one of the biggest advantages of HDTV is the built-in Dolby Digital surround sound. The picture is most important, of course, but surround sound really makes a difference in creating the "illusion of being there" that HDTV can offer.

The biggest thing to look for is vibration. Everything vibrates to a degree, including such structures as walls, ductwork, light fixtures, and woodwork. When your subwoofer belches out a low-frequency sound wave that is absorbed by the room's elements, they vibrate in reaction and establish the room's own special interpretation of the sound wave. Ultimately, the direct sound coming from the speakers combines with the secondary sounds coming from all this vibration around you; that total effect creates your acoustic summary of the action on the screen.

Half the job is simply securing everything around the room and listening for things that add noise. Subwoofers can shake things up too much; get inexpensive isolation pads for the subwoofer's feet. If you're using an HDTV projector, consider a special mounting for it that contains the noise it adds to the equation. (Remember not to block the projector's fan and airflow; it puts off a lot of heat, too.)

Try to avoid showing off your gear. Many enthusiasts feel an urge to put all their equipment out in the open where people can see their investment. This merely adds more noise and heat to your room, which you have to compensate for with cooling fans and more soundproofing. Tuck away as much of this stuff as you can — out of sight, out of sound range (if possible), and away from dust. People are here to see the HDTV, not your shiny new amps!

To get the most from that great digital surround sound your HDTV system is capable of putting out, take a few steps to optimize your HDTV theater room's audio environment. Try to control two major killers of surround-sound goodness:

- Sounds coming into the room from other sources
- Unwanted sound reflections and refractions as the sound waves from your speakers interact with the room

Building from scratch (or doing a major renovation)

If you're building a new home from scratch or doing a renovation, you're lucky because you can optimize your environment at the structural level. You basically can design your own isolation chamber that helps preserve the impact of the HDTV content.

A few steps can optimize your new HDTV space for your theatrical presentations:

- **Build a new room within a room.** A room within a room can isolate and control the impact of the sound system's signals on the room itself. This room would have its own walls, flooring, and ceiling to optimize the HDTV experience.
- **Get some distance.** When building the walls, make sure the studs of two adjacent walls (the new inside one and the existing home walls) don't touch each other. "Decoupling" these studs cuts off a main path for sound to travel into and out of the room.
- **Add more material.** You can reduce unwanted vibrations (and what else is sound?) by adding either (or both)
  - A second layer of sheetrock to your wall
  - More insulation inside the wall
- **Soundproof the walls.** Soundproofing is easy to install between the studs and your drywall. It should reduce both
  - Sounds traveling into and out of the room
  - Vibrations from the drywall against the studs

  Acoustiblok (`www.acoustiblok.com`) has a helpful tarpaper-like sound-barrier material that you just tack onto the studs. Firms like Quiet Solution (`www.quietsolution.com`) make special sheetrock designed for trapping noise.

  Follow the manufacturer's instructions when you apply soundproofing. Most manufacturers tell you to install soundproofing material somewhat *loosely* so vibrations can be dissipated in the material itself.
- **If you can, soundproof the floor.** If you have the luxury of installing an additional floor, it acts to separate the theater environment from the floor, in the same way we just discussed creating a wall within a wall. It's usually three layers:
  - The bottom layer usually is sponge-glued to the floor.
  - The middle layer is plasterboard.
  - The top layer is tongue-and-groove chipboard.

Truly advanced plans call for laying a new floor foundation with the same sort of soundproofing we discussed for the walls. If you can do that, you have a *really* understanding spouse!

If you're putting studs onto a concrete floor, consider adding some isolators that cushion the studs from the concrete, keeping the resonance of the sound within the flooring on "your side" of the room. An example is Acoustic Innovations' IsoBloc Isolators (`www.acousticinnovations.com`), which raise standard studs (such as 2x4, 2x6, and 2x8) a quarter of an inch off the floor for more tactile bass effects with the lower frequencies.

- **Remember the ceiling.** If you have (or plan) a suspended ceiling, check into special *spring mounts* so your ceiling doesn't rattle. Companies such as Kinetics Noise Control (`www.kineticsnoise.com`) have special kits for suspension ceilings to isolate sound.

  If you skip this step, you'll regret it the first time you play any movie involving earthquakes or explosions.

If your HDTV home theater is in a basement with a concrete floor, consider adding a *moisture barrier* coating between the concrete and your flooring surface (such as a rug).

Upgrading your existing room for HDTV

If you aren't rebuilding your home, you can improve your existing infrastructure in a number of ways for a much higher-quality sound experience.

The following ideas apply whether you're building from scratch or just prettying up a room for your HDTV:

- **Apply wall sound control panels.** Excessive reflected sound tends to blur the sound image and mar the intelligibility of voice tracks. With your drywalls complete, you can help control your room's sound image by mounting sound-control panels on the walls to help control refracted and reflected sounds. When you go to theaters, you oftentimes see sound-control panels hanging on the walls in funky geometric shapes. They're made of specially shaped absorptive and reflective materials to better guide and groom the sound in the room. Sound-control treatments cut down on the amount of sound bouncing around the room, enhancing low-level dialogue and environmental effects delivered over today's high-quality audio systems. You typically place these at your speakers' first reflection points (typically the sidewall boundaries and rear wall behind the main listening position).

Speaker *reflection points* are easy to find: Have your spouse or a friend move a mirror along each side wall while you're sitting in the spot where your couch or theater seats will go. When you can see the speaker through the mirror, this is your first reflection point, and a great spot for absorptive material.

Acoustic Innovations (www.acousticinnovations.com) has hardwood frames in mahogany, cherry, walnut, and oak stains for its Maestro line of panels. Kinetics Noise Control (www.kineticsnoise.com) offers a Home Theater Absorption Kit that also contains special midwall decorative and absorptive panels that take care of sound reverberation.

Consider the corners of the room. Kinetics Noise Control also makes special triangular corner panels engineered to absorb the low-frequency bass sound that tends to gather in corners. Acoustic Sciences (www.acousticsciences.com) has an attractive product called the Acoustical Soffit, which plays the dual role: It helps control low-end bass response while it hides wiring inside a hidden raceway. If you don't have space to run wiring, you can kill two birds with one stone with products such as the Acoustical Soffit.

- **Remember your open spaces.** If you have open areas in your space such as a big entryway on one side of the room or large glass windows or mirrors, consider covering them with heavy sound-absorbing drapery — this trick can help even out the sound field in the room. You can make your own draperies by using heavy velvet or other drapery material from your local crafts shop, or you can buy some online at any of the home-theater supply sites. In a pinch, any drapes are better than nothing.

Other companies also offer lines of acoustical suppression and enhancement products. Check offerings from companies such as American Micro Industries (www.soundproofoam.com), Auralex (www.auralex.com), GIK Acoustics (www.gikacoustics.com), and Owens Corning (www.owenscorning.com).

If you want acoustical panels on the ceilings or walls, consider using Rotofast Panel Anchors (www.rotofast.com) to really secure the panels to walls. If you want to know some other tips/tricks for acoustical room building, check out Acoustics Forum (forum.studiotips.com). Although you can find lots of topic-specific groups within this forum, most of the commentary that's any good is the back and forth occurring in the general discussion forum.

You can spend as much as you want to soundproof your HDTV environs, from a couple hundred dollars to more than $50,000. You should add at least some soundproofing, especially if this is a new space without much else in the room. You can find some very reasonable soundproofing treatment packages for your room; they also add a professional look to the finished room.

Shining the Right Light

As with the sound in your room, controlling lighting is critical to maintaining the right atmosphere for your HDTV environment. Too much here, too little there, stray light on one side of the room — any and all of it is noticeable when you start using your HDTV enough.

You can spend as much or as little as you want on home-theater lighting. We cover $5 dimmer switches and the $300 lighting systems, but what you do is merely a reflection (no pun intended) of what you want to spend on your HDTV viewing room.

We'll warn you right now that adding lighting to your room is the type of project that you can get carried away with. Unlike sound treatments, lighting can get fancy and it can be fun — like your own preshow, er, show! (See the nearby sidebar, "The stars at night are big and bright . . ." to find out about fiber optic starfields on the ceiling.)

"The stars at night are big and bright . . ."

You don't have to be deep in the heart of Texas to see stars on your HDTV room's ceiling. In fact, you can buy neat fiber optic kits to add all sorts of effects to your HDTV theater. Companies such as Acoustic Innovations (`www.acousticinnovations.com`) offer fiber optic kits to

- **Cover your ceiling:** To make your ceiling look like a nighttime sky, you can apply fiber optic starfield ceiling panels to your existing ceiling. The panels both make your ceiling more fun and offer acoustical correction.
- **Cover your walls:** You can install infinite starfield panels that are made of multiple polycarbonate mirrors and fiber optic lighting and create an illusion of infinite depth. Pretty neat if you're into psychedelic midnight shows!
- **Cover your floor:** You can get fiber optic carpeting that's filled with tiny points of light (50 fiber optic points per square foot, to be precise).
- **Cover your windows:** You can get classic, velvety, fiber optic curtains that can twinkle, too.

Heck, if you put all these in one room, you can probably think about charging admission for the room itself, much less the movie!

Each of the fiber optic kits comes with a dimmer, speed control for twinkle, and an on/off switch. You even get to choose between a twinkle or color-change wheel. Gosh, if these were the only decisions we had to make in life! Fiber optic starfield lights run about $50 per unit.

For $5, you can get a dimmer switch from Lowes, Sears Hardware, or Al's Corner Drugstore. If your room has a central light, you can create the same low-light theater you encounter at the cinema; you can see where you're putting your soda and popcorn, but keep the aura of a theater. We think this is a no-brainer decision if you do nothing else.

For not a heck of a lot more, though, usually around $200 to $300, a lighting-control system can give you control of lights within your HDTV room (and the rest of the house). These have a wall-mounted keypad, a remote control, or both; you can turn on, turn off, brighten, dim, or otherwise control each light in the room. Some packages use radio-frequency signals to communicate with the light jacks; others use X10 (see the sidebar "What's X10?") or other newer home control signaling options.

Although finding dimmer systems that work with halogen or fluorescent lights isn't impossible, any dimming light control works with standard *incandescent* bulbs.

Why would you want anything more elaborate than a dimmer switch? Well, think about this scenario: When your guests arrive to watch a movie or the NBA finals in HDTV, your lighting is in Arrival mode (house lights are up, the bar is brightly lit, accent lights are on). When the show is ready to begin, you hit a button to enter a flashing-light period, and then lighting highlights the seating and walking paths. When you're ready to start watching the movie, another button dims the lighting altogether while highlighting the screen — then it fades, the curtains part, and the show starts.

We talk more about how to do things like this in *Smart Homes For Dummies* and *Home Theater For Dummies,* but without too much effort or construction, you can add some very effective and professional lighting effects to your home.

You can get single-room lighting systems from players such as Leviton, Lightolier Controls (`www.lolcontrols.com`), LiteTouch (`www.litetouch.com`), Lutron (`www.lutron.com`), Powerline Control Systems (`www.pcslighting.com`), and Vantage (`www.vantageinc.com`).

Lutron, for example, has a whole line of products geared for your HDTV theater. Lutron's Maestro IR package is a simple entry-level package that allows you to remotely control a dimmer switch in the room. Its GRAFIK Eye product adds more preset scene options and control of window shades and curtains. Lutron's RadioRA is a wireless, whole-home lighting-control system that uses radio-frequency technology instead of power lines for signal communication. With any of these packages, you can create the lighting scenarios mentioned earlier in this section. These packages range in price from $10 to $1,500.

What's X10?

If you start looking around the Web for info about home automation, you're going to hear about X10. X10 is important if you're talking about installing lighting controls in your home without having to run a lot of new wiring. For years, X10 has been the dominant protocol for controlling (turning on, turning off, and dimming) electrical devices such as lights and appliances — through your home's elec-trical lines. You can find X10 gear from a bunch of companies, including Leviton (www.leviton.com) and X10 Wireless Technologies (www.X10.com).

Using X10 couldn't be simpler. You basically plug in an X10 wall *module* — a small, box-shaped device no bigger than the "wall-wart" AC transformers that power telephones and answering machines — right into the wall outlet. Then the appliance (a lamp, for example) plugs into the module.

X10 *controllers* send their control signals over your power lines to every outlet in the house. When the controller finds the module it wants, it changes that module to the desired state (such as "Off"). The following figure shows a limited X10 network.

However, X10 as a protocol is being overrun by several next-generation technologies. You can find out about INSTEON (www.insteon.net), ZigBee (www.zigbee.org), and Z-Wave (www.z-wavealliance.org) by visiting their Web sites. Most of the major electrical, lighting, and other home-equipment manufacturers are hopping on board one or more of these next-generation options. Check out those sites for some ideas about what packages can power your HDTV room.

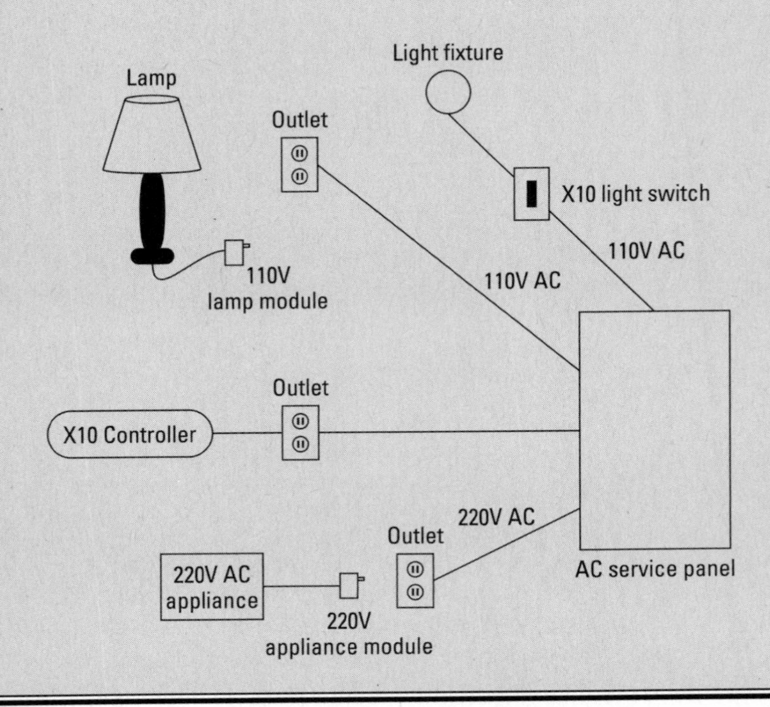

The motorized shade/drape kits for your HDTV screen or windows are simple to install. If you've ever installed a drapery rod, such a kit is basically the same thing with a motor. One kit, the Makita Motorized Drape System ($600, `www.smarthome.com`), is a self-contained system for your HDTV theater drapes. It comes with the tracks and mounting for the system — you just add the drapes. You can also get motorized window-treatment hardware from manufacturers such as BTX (`www.btxinc.com`), Hunter Douglas (`www.hunterdouglas.com`), Lutron (`www.lutron.com/sivoia`), and Somfy (`www.somfysystems.com`).

A great resource for ideas about lighting and room control systems — and your home theater in general — is the magazine *Electronic House* (`www.electronichouse.com`). It's inexpensive, available on newsstands, and has helpful hints about how to approach your projects from lots of different angles (including lots of home-theater-specific articles).

Chapter 20

Enhancing Your HDTV

In This Chapter

- Scoping out factory defaults
- Understanding picture settings
- Calibrating your HDTV
- Leveraging hardware
- Bringing in the pros

When you take your HDTV out of the box and turn it on, it's in what many experts call *torch mode:* The brightness of your picture is cranked way, way up. Manufacturers do this because of the retail environment — lots of bright overhead lights that could wash out the picture on the showroom floor. Manufacturers also crank up the picture brightness because, when 10 or 12 HDTVs are lined up next to each other showing the same video, the brightest picture tends to catch your eye, so you think, "Must buy *that* one!"

In this chapter, we tell you about the different adjustments you can make to your HDTV's picture. We also tell you about a series of different inexpensive DVDs (and a pair of D-VHS tapes) you can buy to really get things set up correctly — just adjusting the picture by eyeballing it isn't going to cut it. We even tell you about some free DVD software that can do most (but not all) of what the other DVDs can do. Finally, we tell you about the ultimate in HDTV maintenance: the professional calibration. It's like a spa day for your TV!

Why You Need to Calibrate

Unfortunately, the torch mode your set is adjusted to when it comes out of the box *can't* give you the best picture quality when that HDTV is in your home. In fact, it gives you a *lousy* picture in a darkly lit room (which is best for movie watching). Even in a normally daytime-lit room, the standard picture settings of most TVs are too bright for best picture quality.

That super-bright mode also shortens your HDTV's lifespan, or at least the lifespan of your light bulb, if you have a microprojector system, which really means something if you have a $10,000 plasma that you were planning on using for 10 or 15 years! Even if you have to change only a bulb (and not replace a tube or buy an entirely new TV), you're still out a couple of hundred dollars.

Getting Your Settings Right

To *calibrate* (properly configure) your HDTV's picture, you need to delve into the TV's menu system, using your remote control and the *on-screen display* (or OSD) that your HDTV provides.

Every HDTV differs, but you most likely end up in a menu called Video Settings to adjust the picture elements we discuss here.

We probably don't own the same HDTV as you do. We've seen/used/played with many different sets — enough to recognize that we can't 100-percent match up to the terms that your HDTV uses in its menu system for picture quality settings. So keep that in mind — we use the generic terms here.

Here are the most common picture settings:

- **Contrast (white level):** The control adjusts the *white level,* or the amount of whiteness your screen displays:
  - If your white level is too high, white areas of your picture tend to bleed over into the darker areas that surround them.
  - If your white level is too low, whites don't appear true white.

  Whites and blacks are measured on a scale called IRE (Institute of Radio Engineers) units. These units are percentages between 0 and 100 — 0 percent is black; 100 percent is white.

- **Brightness (black level):** How's this for confusing? Your HDTV's *brightness* control adjusts the *black level* that you see on-screen. Seems a bit counterintuitive, doesn't it? If the black level is set incorrectly, dark scenes on your HDTV are indistinguishable — you won't see the bad guy in the black suit hiding in the shadows.
- **Sharpness:** Ever see a fuzzy picture on your screen? The sharpness control is the likely cause — it adjusts the *fine detail* of the picture.
  - If the sharpness is set too high, your picture appears *edgy,* often with blobs around the edges of objects, instead of clearly defined lines.
  - If the sharpness is set too low, you have a fuzzy picture. (Aha, so *that's* why it's fuzzy!)

- **Balancing:** Two adjustments set the balance of colors:
  - *Color:* The color setting adjusts the *intensity* of the TV's display of colors — if this is set too low, you see only black and white (or grayscale); set it too high, and your colors bleed together.

    If your color setting is too low, images begin to appear as black and white. If the color level is too high, images take on a reddish tinge — everyone resembles Bozo the Clown!
  - *Tint (hue):* On most TVs, this control is labeled *tint,* but a few are technically correct and call it *hue.* This control adjusts your display's color only within the range between red and green.

    This is pretty hard to adjust with your eyes alone — we recommend the calibration systems in the next section to get it done right.

Many HDTVs come with some picture presettings — mix-and-match combinations of the settings listed here — designed for specific purposes. For example, Sony HDTVs have settings such as Vivid (which is that showroom torch setting), and Pro (which often is pretty close to being well calibrated for viewing movies in a dark room).

Getting Calibrated

All those video settings discussed in the previous section can be a bit difficult (we think impossible) to set up properly by just eyeballing it. Unless you have the visual equivalent of perfect pitch, how can you know when the whites are white enough and the blacks are black enough (not to mention whether the reds are red enough)? You can either do it yourself or bring in the professionals.

Doing it yourself

Calibration DVDs and tapes contain a specially designed series of tests and test patterns that you can follow (along with on-screen directions) to get the settings right on your HDTV. Calibration systems provide all the information you need to tune your HDTV.

If you really have to adjust your video without the aid of a calibration system, here's what we recommend (at minimum):

- Turn down the brightness and contrast until the levels are about ⅓ to ½, all across your screen.
- Substantially turn down the sharpness control.

Commercial DVDs

Most calibration systems come in the form of DVDs, which means you won't be using a high-definition picture to tune your HDTV. Yes, this is a less-than-great situation, but the settings are fairly universal, so an HDTV that's been well calibrated for a DVD input also looks good for displaying HDTV.

High-definition DVDs are here (see Chapter 11). Expect to see high-def versions of the calibration DVDs we've mentioned here before long.

Popular calibration DVDs are on the market from a couple of sources. Any of these discs do a great job helping you get your HDTV tuned up and calibrated. Each of these discs includes

- A funky blue filter — they usually look like those crazy, paper, 3D glasses you needed to wear for *Jaws 3!* The blue filter lets you properly set up your color and hue settings.
- Audio-calibration tools that help you get your surround-sound audio system properly configured (if you have one — see Chapter 18 to find out about surround sound).

You really can't go wrong with any of these discs:

- **Avia: Guide to Home Theater:** Available from Ovation Multimedia (`www.ovationmultimedia.com`), this disc contains a ton of great background material about HDTV and home theater. It explains a lot of the same stuff we talk about in this book. It also contains a series of easy-to-follow, on-screen test patterns and signals that let you correctly adjust all the settings we discuss in the preceding section. You even find test tones to help calibrate your surround-sound audio systems. This disc runs $49.95 directly from Ovation.

  Most folks probably find it to be overkill, but Ovation also produces a professional-quality variant of its test disc, the Avia PRO. This $400 set consists of seven DVDs and an elaborate user manual for the ultimate in calibration.
- **Digital Video Essentials:** Found online at `www.videoessentials.com`, this is the definitive calibration disc. One really cool thing about Video Essentials is the inclusion of video footage that you can watch to see the results of your adjustments with actual video, instead of just on a test pattern. Digital Video Essentials lists for $24.95 and is widely available online.

  The folks at Joe Kane Productions (who created Digital Video Essentials) also have a six-disc professional set called *Digital Video Essentials Professional,* which retails for $350 and includes some high-definition calibrations, using the Windows Media system for high-definition 720p and 1080i signals. Like the Avia PRO, this is probably overkill for most folks (you can probably have a professional calibrate your display for less money), but we wanted to let you know this is out there.

- ***Sound & Vision*** **Home Theater Tune-Up:** This one is also produced by Ovation Software, but in conjunction with *Sound & Vision* magazine (one of our favorites). In addition to general display calibration, this disc includes tests that demonstrate aspect ratios and let you test the S-video and component-video outputs on your DVD player to see which works better in your system. You can find the disc for only $21.95 at `http://shop.soundandvisionmag.com`.

Using any of these discs is a great way to calibrate your HDTV. Most people find the Sound & Vision disc to be the easiest, but we think all three are simple enough to use if you just follow the instructions on-screen.

THX DVD

If you don't want to spend the money on a calibration disc, you can get much of the functionality without spending an extra penny. Just pick up a movie DVD with the *THX Optimizer* on board (there are thousands). Like the commercial calibration discs, the Optimizer walks you through a series of steps to adjust your display (and your audio system). Any THX certified movie (you'll see the THX logo on the front) includes the Optimizer.

To get all you can from the Optimizer, you need a blue filter. You can get one from THX by filling out the order form online. (A link is on the front page of `www.thx.com`.) All you need to pay is two bucks for shipping and handling.

High-definition D-VHS

If you're one of the few folks with a D-VHS player (they're hard to find these days), the Video Essentials folks (`www.videoessentials.com`) have responded to your needs by coming up with D-VHS videocassettes that let you calibrate your HDTV with a true HDTV picture, instead of an NTSC-quality DVD picture. These tapes are called *Digital Video Essentials — High Definition.*

You need a D-VHS VCR to use one of these tapes!

Digital Video Essentials — High Definition is available in two resolutions. Either costs about $90. Choose the version that's closest to your TV's native resolution:

- If you have a direct-view CRT or a CRT-projection HDTV system, you probably need the *1080i* version. You'll also want to use this version if your LCD, DLP, or LCoS system has a 1080p native resolution.
- If your HDTV is a 720p or 768p plasma, DLP, or LCD system, you probably need the *720p* version.

Going beyond the eyeball: Using calibration hardware

If you want to get really fancy with your home-theater calibration process, you might invest in an *optical comparator* or a *colorimeter.* Optical comparators are simply electronic devices that can read the colors and brightness of your screen and provide you with an accurate way of calibrating your picture — they're what the pros use when they come to your home and set up your TV. (We discuss professional calibrations in the next section.)

Using an optical comparator or colorimeter can be especially useful if you're one of those folks with some degree of color blindness (which can include, depending on which study you believe, up to 1 in 12 men).

Most optical comparators are too expensive for the average person — they can cost as much as a 50-inch plasma HDTV! For example, Ovation Multimedia's OpticONE system, a popular hardware and software solution, retails for $2,200.

But there *are* solutions that regular folks can afford, like ColorVision's Datacolor SpyderTV ($229, `www.colorvision.com`). The name comes from the fact that it has spider-like suction-cup legs that attach it to your TV — ignoring the fact that it has three legs, not eight!

To use the SpyderTV, you simply stick the unit's suction cups to the front of your display and then plug the USB cable running out of the SpyderTV into your computer. (Windows 2000 or XP is required, and a laptop is a heck of a lot more convenient for this task but isn't required.)

For flat-panel plasma and LCD displays, the SpyderTV includes a special suction-cup attachment device — you're not supposed to use the suction cups built into the three legs of the SpyderTV for these kinds of displays.

After you hook up the SpyderTV, just put the supplied DVD in your DVD player and run the supplied software on your PC. Follow the instructions on your PC screen, using the HDTV's remote control to adjust the picture. The SpyderTV tracks your changes and gives you feedback when you have the right settings.

Bringing in the pros

The absolute best way to get your HDTV properly tuned up is to hire a pro to come to your home and do a professional calibration. Two things separate a professional calibration from the one you perform yourself at home:

- **Training:** Your calibration professional should have extensive training and certification, along with a chunk of real-world experience with finicky TVs (things you probably don't have).
- **Equipment:** A professional also has some expensive equipment that can measure the color and brightness of images on your TV screen — this gives a much more precise calibration than using your eyeballs.

A professional calibration usually costs between $200 and $500, depending on what type of display you have.

If you have a professional calibration done, make sure you choose someone who has been certified and trained by the *Imaging Science Foundation,* or ISF. The ISF has trained (and continues to train) thousands of home-theater dealers in the art of system calibration. You can find a trained calibration professional near you by searching ISF's Web site at `www.imagingscience.com`.

Don't call ISF directly to ask someone to calibrate your system. They'd know how, but that's not what they do for a living. The Web site includes a searchable (by state) listing of the dealers they've trained to perform this service. (Training and certifying folks who do calibrations *is* what they do!)

Part VI
Geek Stuff

"The picture is so sharp, you can see the Hot Wheels logo on the cars that the monster is throwing across the city."

In this part . . .

HDTV technology is, depending on your enthusiasm, either enthralling or mind-numbing. The chapters in this part lay out exactly how the pictures get on the screen. We explore the details of HD television engineering that will allow us ultimately to compare your HDTV set options. We make sure you're well founded in concepts like interlacing and 3:2 pulldown correction (and no, that's not a pro wrestling move). We answer the most common questions we get asked about HDTV so you can look sooooo smart at your next party. "Why yes, I can tell you the difference between an HDTV and HDTV-ready television!" "Oh, you're sooooo smart!"

Then we talk all about the different options you have for which great TV set to choose. We talk about front- and rear-projection systems, plasma and LCD flat panels, and CRT-powered TV sets — and the pros and cons of each approach. Despite what the salesperson in your local electronics store has told you, "everyone" does not buy the same model, much less the same type of HDTV. Different types are suited for different applications, and we help you figure that out in this part.

By the time you're done with this part, you'll appreciate each HDTV type and want one of each. As they say on *Star Trek Next Generation,* "You have been assimilated!"

Chapter 21

TV Engineering

In This Chapter

- Learning lines, picking pixels
- Resolving your picture
- Fixing your pixels
- Scanning away
- Doing the 3:2 pulldown

In Chapter 1, we give you a 20,000-foot, fly-by view of HDTV (and TV technology in general). The deeper you get into the HDTV world, the more complex and complicated the whole technical shebang gets; you go from HDTV versus analog into arcane discussions of lines of resolution, pixels, scanning systems, fixed-pixel displays, scaling systems, and more.

It's enough to make your head spin, trust us. So, in this chapter, we give you the deeper details that we glossed over in Chapter 1. We explain the similarities (and differences) between lines and pixels, and we give you a firm grasp of horizontal versus vertical. (It's not as easy as it sounds, believe it or not.) We also provide more details about interlaced and progressive-scan pictures, and we give you a primer on frames, fields, and film versus video. It's fun, so read on!

Lines and Pixels

Okay, we admit it. When we first got into the whole video/TV/HDTV thing, we thought we had a good grip on the concept of lines of resolution (or *scan lines*), pixels, and any other metric used to describe *resolution.* Resolution is the level of clarity and fine picture detail on your screen.

But we were wrong — all the horizontal this and vertical that and lines and pixels were just too much to keep up with at times. Add to this the fact that *horizontal* resolutions can sometimes refer to lines that run *vertically,* and vice versa. It's a big mess. We hope that the following discussion makes it clear in your head. Just remember the shape of a TV screen (it's wider than it is high, and that the horizontal resolution is the larger number — across the width of the TV) whereas the vertical resolution measures the resolution of the height of the TV. The vertical resolution number is the one you hear most often (for example, 720p or 1080i).

Going with lines

The oldest, most common way of describing a TV's resolution comes from the world of CRT (cathode-ray tubes; see Chapter 24) TVs. These TVs create an image by shooting an electron beam at the screen and moving *(scanning)* from left to right and top to bottom. The lines that result are referred to as *scan lines.*

We're talking about the capabilities of the TV itself here — program images have their own resolutions. For example, an HDTV program might be 720p or 1080i — numbers that refer to the total scan lines contained in the program material.

A TV's *vertical resolution* (the metric most commonly referred to) is the number of lines that the TV can display in the vertical direction. Think of a stack of pancakes, with each pancake representing a line moving horizontally — yes, it's confusing — across the screen. The total number of pancakes in the stack (lines on-screen) that the TV can display is its vertical resolution. See Figure 21-1 for a picture that represents this.

Don't forget! Vertical resolution measures the number of lines that run horizontally across your screen!

We're sort of breaking the rules by describing scan lines and vertical lines of resolution as being the same thing. They're related, they're similar, they look/feel/smell alike, but *technically* scan lines are the lines within a TV program, and vertical resolution refers to how many lines the TV's hardware can show. We think it's a lot easier to envision if you think of the electron gun scanning across the screen first, so we presented it that way.

If a TV can't display at least 720 vertical lines of resolution (720 delicious pancakes), it isn't an HDTV-capable TV.

Now take a single one of those lines/pancakes and look across it horizontally. Each TV can display a limited number of individual picture elements (for example, dots of alternating colors) along that line and still keep it legible. This limit is the *horizontal resolution* of the TV. Figure 21-2 demonstrates horizontal resolution.

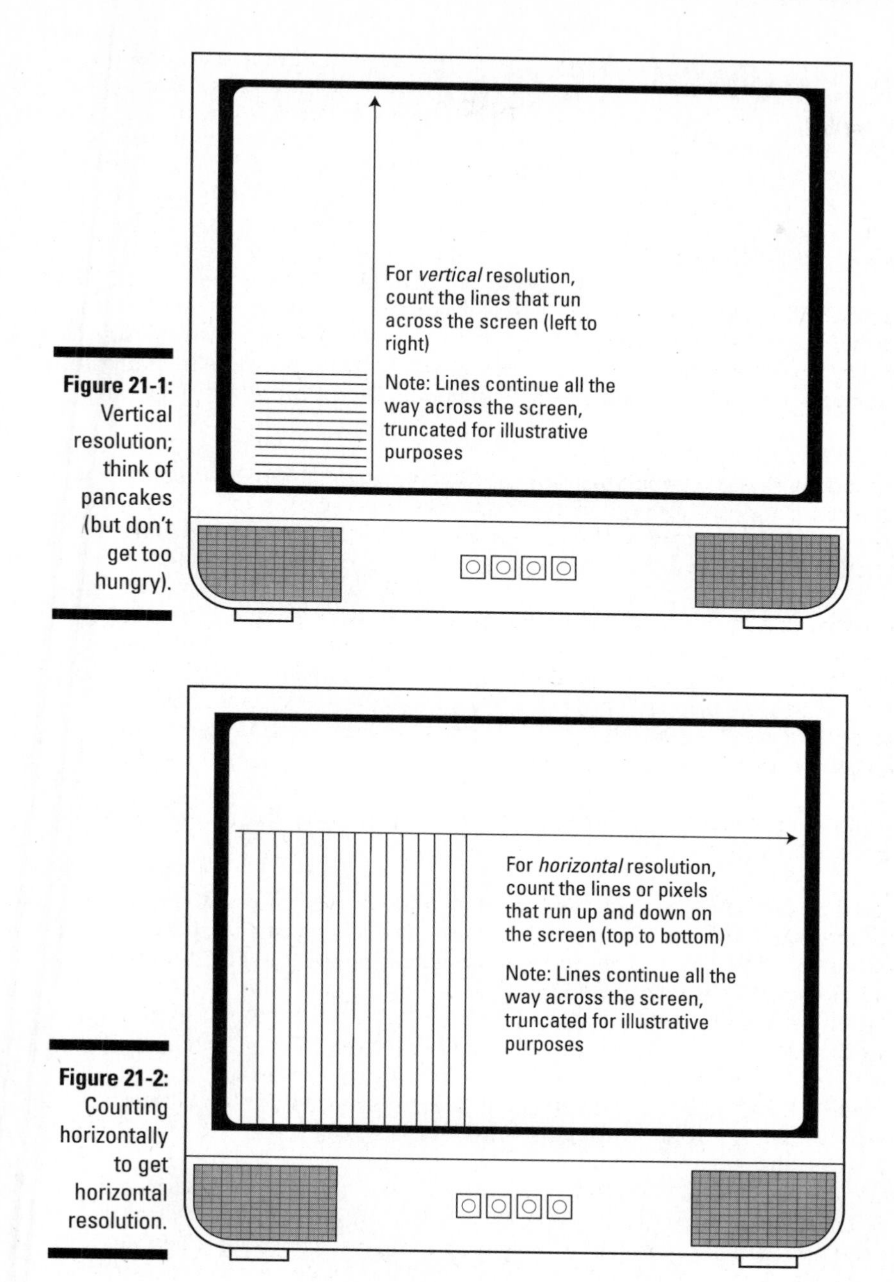

Figure 21-1: Vertical resolution; think of pancakes (but don't get too hungry).

Figure 21-2: Counting horizontally to get horizontal resolution.

Picking pixels

Although scanning and lines are the gist of what happens in an old-fashioned CRT TV (even a fancy HDTV version of a CRT TV), many newer TV technologies are *fixed-pixel* displays. What this means is that these TVs, by nature of their physical design, don't have an electron beam moving across a screen, but instead have thousands (or millions) of individual picture elements *(pixels)* that light up to create the picture. These displays are "fixed" because (unlike a CRT) they won't let you change the number of pixels in the display by re-aiming or refocusing an electron beam. Another happy feature of the fixed-pixel display is that each pixel is electronically addressed with exacting light and color value and not subject to magnetic distortion, as the CRT can be.

LCD and plasma flat-panel TVs (Chapter 23) and LCD, LCoS, and DLP projection TVs (Chapter 22) are all fixed-pixel displays. In a fixed-pixel display, resolution is determined by simply counting the number of pixels in a vertical stack (from top to bottom or vice versa) for the vertical resolution and across the screen (left to right or vice versa) for horizontal resolution.

The resolution of a fixed-pixel display is usually written or referenced as "horizontal resolution x vertical resolution." For example, Pat's Panasonic plasma HDTV resolution is 1,366 x 768 pixels — 1366 pixels in the horizontal direction, and 768 vertically.

If you're used to setting up the resolution of a computer screen, you already have this concept down. A screen resolution of 1,024 x 768 (the most common PC-monitor resolution setting) is just 1,024 horizontal pixels and 768 vertical pixels.

As with the lines of resolution, it's the vertical number of pixels that is most commonly referred to, and which is essential when deciding whether a TV can be called an HDTV. A fixed-pixel display must have at least 720 vertical pixels to be called HDTV-capable.

The horizontal and vertical resolution in a fixed-pixel display is that display's *native resolution.* A TV signal (whether from broadcast or a recording) must be converted to this native resolution to be displayed on the fixed-pixel screen. The device that performs this task is the *scaler,* which we describe in detail in Chapter 16.

Defining standard and high resolutions

Knowing all these lines of resolutions and numbers of pixels can be an interesting intellectual exercise (if you simply *gotta* know all the stats of your HDTV), but you really need some sort of frame of reference to understand what's good and what's not.

The magic million (or so)

Instead of just comparing vertical (or horizontal) lines of resolution, it can be very instructive to compare HDTV to other TV sources (such as DVD or analog NTSC broadcasts) by comparing *total* pixels. That is to say, by doing the math, and multiplying horizontal by vertical.

A typical analog NTSC broadcast, for example, might show 330 horizontal pixels by 480 vertical pixels, for a total 158,400 pixels. An anamorphic widescreen DVD zooms up to 345,600 pixels. But even this number pales next to the potential of over *two million* pixels in a 1080i HDTV picture. Even the lower resolution 720p HDTV picture hits 921,600 pixels.

If you haven't seen it yet, you'll be simply amazed at what a million (give or take) pixels on-screen means! It's a huge difference from even the best DVD picture.

Remember that pixels aren't the only determinant of picture quality; things like screen size (and therefore how close together the pixels are), pixel shape, color accuracy, brightness, and more all work together to determine picture quality. But resolution is perhaps the most important factor, and HDTV really rules when it comes to res!

We've already mentioned that 720 *vertical* lines of resolution (or pixels) is the baseline requirement to reaching HDTV nirvana, but you probably want to know more details (otherwise, you'd have skipped this techie chapter, right?).

Table 21-1 shows the most common types of TV signals and the TV resolutions required to show them in full detail.

If you're not familiar with the *i* and *p* suffixes in Table 21-1, we explain them in the following section, "Scanning and Interlacing."

Table 21-1 TV-Signal Resolution Requirements

TV Signal Type	*Horizontal Resolution*	*Vertical Resolution*	*Pixels*
NTSC (analog broadcast)	330	480	158,400
NTSC (DVD)	720	480	345,600
SDTV (480i)	640	480	307,200
EDTV (480p)	852	480	408,960
HDTV (720p)	1,280	720	921,600
HDTV (1080i/p)	1,920	1,080	2,073,600

Vertical resolution gets all the attention, but horizontal resolution is important too. (It's not like your eyes can't see sideways as well as they can up and down, right?) The whole HDTV industry, however, is a bit looser with horizontal resolution than it is with vertical. Relatively few HDTVs can reproduce the full 1,920 horizontal pixels/lines that 1080i/p HDTV signals can reproduce, so horizontal resolution is often de-emphasized in marketing materials and in articles about HDTV. You *do* want as much horizontal resolution as you can get, but you shouldn't dismiss an HDTV just because it can't reproduce 1,920 horizontal lines of resolution. Add to this the fact that most of the HDTV cameras of today support only 1,440 horizontal lines (to save on recording bandwidth), and you may rightly conclude that a full 1,920 is not be absolutely critical.

Scanning and Interlacing

Resolution doesn't fully define and explain HDTV. Scan type also defines HDTV signals (and HDTV TV hardware requirements). Traditional analog CRT monitors, designed for NTSC programming, use *interlaced* scanning; many (more modern) HDTV designs — including many CRT-based HDTVs — use a system called *progressive scanning.*

Fixed-pixel displays are inherently progressive — that's just the way they work. If an interlaced signal is fed into a fixed-pixel display, the display uses its internal scaler (Chapter 16) to convert this signal to a progressive scan.

Fields, frames, and your TV

Before we get into details about interlaced versus progressive scan, you need to understand two key concepts:

- **Frames:** Frames are complete, full-screen images that make up an instantaneous portion of the HDTV picture. In other words, in a 720p HDTV signal, a frame is all 720 lines of the picture. NTSC images are transmitted at 30 frames per second, whereas HDTV images are transmitted at either 30 or 60 frames per second.
- **Fields:** Fields are simply *half* of a frame. The first field within a frame consists of the odd-numbered scan lines (1, 3, 5, and so on), and the second field consists of the even-numbered scan lines.

When a TV signal (whether it's NTSC or HDTV) is transmitted one field at a time, that signal is *interlaced.* Interlaced TV video has 60 fields per second (actually, 59.94 fields per second, but everyone rounds up!). These 60 fields are equivalent to 30 frames per second (fps). Figure 21-3 shows how a frame is divided into fields.

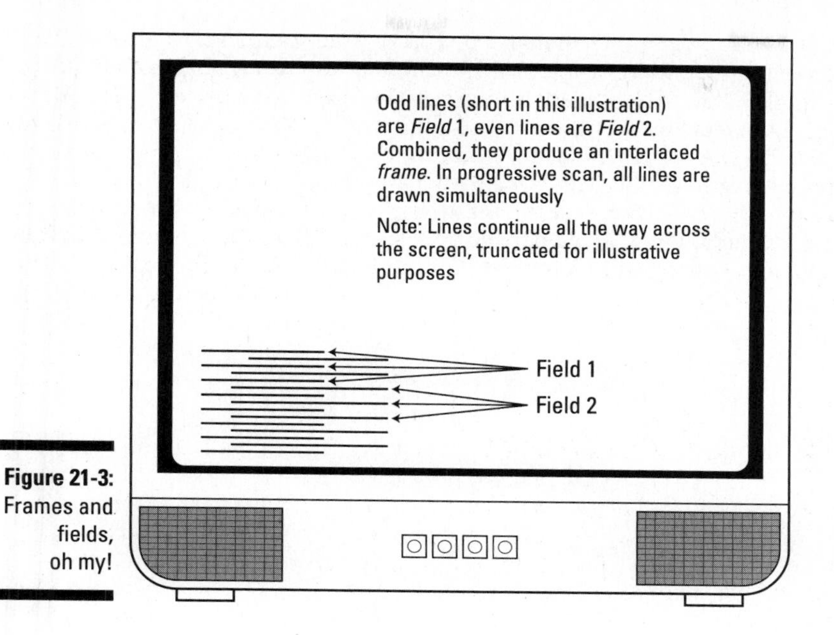

Figure 21-3: Frames and fields, oh my!

In a progressive-scan signal, both fields that make up a frame are sent simultaneously, allowing the TV to create the entire frame at once. In an interlaced system, only half of the picture is "drawn" on the screen at once, but the short time period between fields (60 fields per second!) means that your eyes see the two fields as one.

So what's better, interlace or progressive? Well, first we should answer by saying that there's nothing really wrong with interlaced video — it's not an inherently bad thing. Having said that, progressive-scan video has a smoother, more natural appearance. Interlaced video tends to have a slightly flickery appearance, at least when compared head to head with progressive-scan video.

Doing the pulldown

ATSC TV signals (standard and high definition) operate at either 30 or 60 fps. The film cameras that most movies (and some TV shows) are recorded on operate at 24 fps.

These differing frame rates make it difficult to convert movies to DVDs (or other formats) for standard-definition television viewing — there's not an easy way to evenly divide 24 and 30. The answer is to create an interlaced video recording (such as a DVD) that alternately repeats half the frames of video as 3 *fields* and the others as 2. This process, called *3:2 pulldown,* turns the 24 film frames into 30 TV video frames.

3:2 pulldown is a clever bit of math, but it can lead to *artifacts* (flaws in the image) because sometimes two fields from different frames are combined into a single frame. These frames might not be identical, so you might end up with a jagged picture.

Why are we talking about this? Mainly because many HDTVs include a circuit called something like "reverse 3:2 pulldown" or "3:2 pulldown processing." This circuit examines a TV signal and recognizes 3:2 pulldown in action. Then it does some wizardry and removes the artifacts that 3:2 pulldown creates.

Chapter 22

Projecting a Good Image

In This Chapter

- Understanding projector basics
- Fronting and rearing
- Going with CRT
- Getting into microdisplays
- Hanging your silver screen

When most people think of HDTV, they think (in a Homer Simpsonesque interior-monologue voice), "Mmmmmmmmm, big screens." And when most people think big screens, they think of *projection* TV systems. Projection TVs offer the most bang for the buck in the HDTV world (and in the TV world in general) — the biggest screen for the fewest bucks.

In the past, projection TVs (or at least most of them) also offered a lousy picture. But fear not, those days are behind you now. Projection TVs can also offer world-class HDTV pictures to go along with their large screens and relatively low price tags.

In this chapter, we talk about the two main kinds of projection systems: front-projection systems, which use a separate screen, and all-in-one rear-projection TVs. We get into the underlying technology — the many different systems used to actually project the picture in a projection TV, each with pros and cons. Finally, we give you the lowdown on screen systems for front-projection TVs.

Projection HDTV Design

Projection HDTVs have two distinct formats:

- Rear-projection systems encapsulate the projection system and the screen into a single chassis. These systems beam the TV picture onto the *back* of the screen.
- Front-projection systems consist of two pieces: the projector itself and a separate screen. A front-projection system beams the image onto the *front* side of the screen.

Rear projection

Most projection HDTVs sold today are rear-projection TVs (RPTVs). When compared with front-projection systems, the biggest advantage of the RPTV is simplicity — all the pieces and parts are in one chassis, just as they are in an old-fashioned CRT (cathode-ray tube) TV. (See Chapter 24 for more detail on these TVs.) So with an RPTV, there's no lens adjusting and focusing of the picture on the screen. In fact, with most modern RPTV systems (such as DLP and LCD RPTVs, which we discuss in the "Projection TV Systems" section), you don't even need to align or aim the picture to avoid those awful ghosty images you might have seen on older RPTVs. Figure 22-1 shows a typical RPTV.

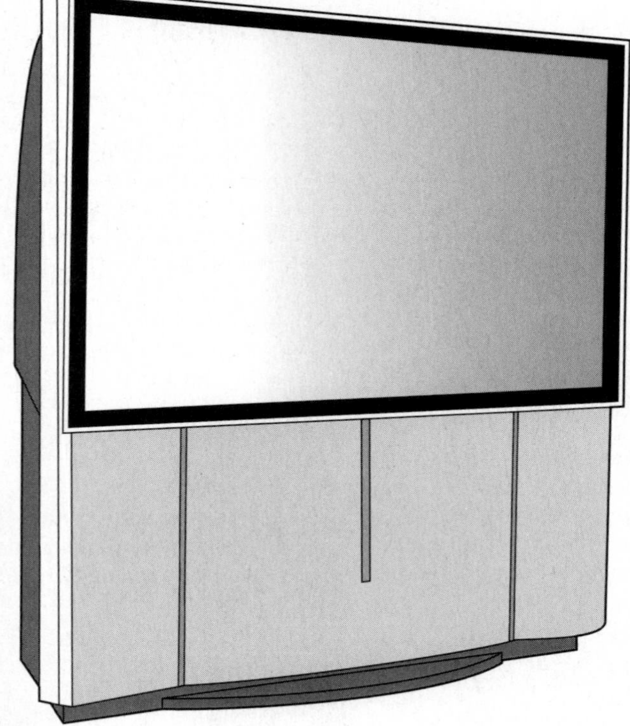

Figure 22-1: The high-def, big-screen bargain: the RPTV.

Compared with a front-projection HDTV, an RPTV

- **Has a large (but relatively smaller) screen:** RPTVs typically range from 42 to 70 inches diagonally (they can get bigger, but those are the most common sizes), whereas front-projection systems can fill screens 100 inches or larger.
- **Is generally less expensive:** The most expensive RPTVs cost more than entry-level front projectors, but you can pay $2,000 or less for an entry-level HDTV RPTV, whereas most HD-capable front projectors start at about $4,000 or more.
- **Is easier to install and set up:** For most RPTVs, you simply plug it in, turn it on, and that's it. Even the easiest-to-install front-projection system requires adjustment in the placement of lens and chassis to focus the picture to the right size on your screen.

As far as picture quality is concerned, RPTVs (just like front projectors) can have extraordinary picture quality. The biggest factor, when it comes to picture quality, isn't so much RPTV versus front projector as it is the *type* of projection system — and the quality of the individual projector.

The biggest picture shortcoming with most RPTVs revolves around the *viewing angle* of the RPTV — how far from perpendicular to the screen a viewer can be and still see the picture clearly. Some RPTVs have poor viewing angles, so they're less than best when viewers are seated far to either side of the HDTV. Figure 22-2 demonstrates viewing angle.

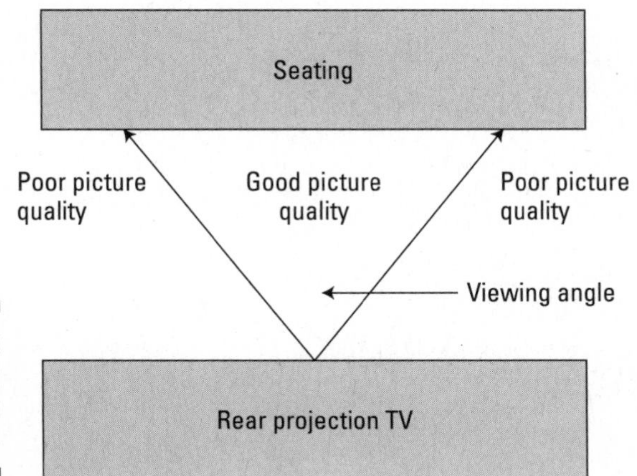

Figure 22-2: The viewing angle of an RPTV.

The other potential shortcoming of RPTVs is their size — not in terms of screen size, but rather in terms of bulkiness. RPTVs based on traditional CRT technologies can be humongous — taking more of your room than you might be willing to allow.

Most new RPTVs, which use microdisplay technologies such as DLP, are actually quite slim — barely thicker than flat-panel TVs such as plasmas. You can even find some that are wall-mountable like a plasma or LCD!

We think that these microdisplay RPTVs are perhaps the best bang for your buck in HDTVs, combining a large screen with a slim overall package, and a picture as good as plasma at a lower price. This is particularly true when you start looking at screens of larger sizes (more than 50 inches).

Front projection

When you really have to have a big, BIG screen for your HDTV, you have only one choice: a front-projection system. A front-projection system is really a lot like being at the movies because you see a (possibly silver) screen in front of you, and a projector is situated behind you. In fact, some movie theaters have begun to give up film and install DLP (digital light processing) projectors that are really just souped-up versions of the same projectors you can buy for your home HDTV use.

Figure 22-3 shows a front-projection projector.

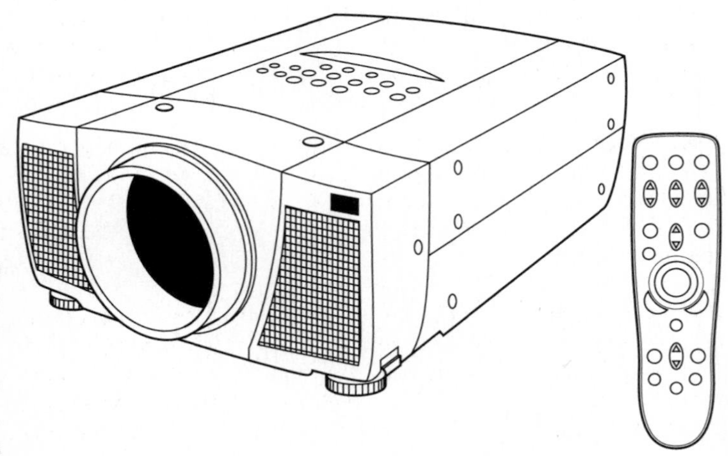

Figure 22-3: Bring the movie theater to your home.

Because you're projecting the image across the room and onto a separate screen, theoretically there's almost no limit to the size of your screen and picture. (Check out Chapter 26, where we talk about projecting a 10-x-20-foot image outdoors!) The only real limit revolves around the brightness of the image put out by the projector. Because a projector must be farther away from a bigger screen (and because it is lighting up a larger physical area), it needs to put out a brighter picture than it would with a smaller screen. Projector brightness is measured using a unit called *ANSI lumens.* Projectors with 500 or so lumens are best on smaller screens or in darker rooms, whereas projectors with 1,000 or more lumens are considered quite bright. Some projectors (usually those designed for PC use) are *really* bright, throwing out 2,000 or more ANSI lumens. These super-bright projectors have been designed for use in brightly lit rooms (like a conference room at an office), but they can do double duty as projectors for very large screens.

Many front-projection TVs have no built-in TV tuner — neither HDTV/ATSC nor standard-definition/NTSC. So you need to use an external tuner, set-top box, or satellite receiver to watch TV.

Projection TV Systems

When you're evaluating projection TV systems, you not only must choose between front and rear projection, you also much choose a projection technology. Projection systems have two types:

- **Traditional projectors based on CRT (cathode-ray tube) projectors:** These CRTs are specialized versions of the same tubes found in traditional TVs to generate the light projected onto the screen. These types of TVs are still available on the market, but generally speaking have been made obsolete by their microdisplay rivals.
- **Microdisplay projectors:** These use microelectronic systems, such as LCDs and DLP chips, in conjunction with a powerful light bulb to project images on the screen. When you go to Best Buy or Sears or Costco and look at the rows and rows of projection TV systems, the majority use a microdisplay technology.

Just a few years ago, we would have told you that CRT projection systems offered the ultimate in picture quality, particularly the very expensive, very top-of-the-line front-projection systems. Nowadays, that isn't quite so true. CRTs are still the best at *black levels,* displaying nuances in very darkly lit scenes and showing a truly black picture on-screen. However, this advantage is diminishing with each new generation of microdisplays. Now that 1080p microdisplays are widely available, CRTs have no other real advantage. Microdisplay projectors trade off a little bit in ultimate picture quality (just a bit, though, related to the black level), but gain in ease of use (pull 'em out of the box and turn 'em on), brightness, and compactness (saving floor space in your viewing room).

We think that for the vast majority of folks, microdisplays are the way to go when it comes to projection HDTVs. If you're looking for the biggest screen on the smallest budget, however, and you have room for a bigger (deeper) cabinet, you can get a great deal on a CRT rear-projection TV (with 52-inch units being sold for well under $1,000 at big retailers like Wal-Mart).

CRT

The granddaddy of projection TV systems is the CRT projection TV. The tubes in these projection TVs are similar to those used in direct-view tube televisions (discussed in Chapter 24 — these are the regular old TVs), but with one major difference: Whereas a direct-view HDTV has a single tube, a projection CRT HDTV has three tubes, one each for red, green, and blue colors.

Available in both rear- and front-projection systems, the CRT is simultaneously a pain in the neck and potentially the highest-quality projection TV you can buy. The three tubes account for the pain-in-the-neck quality — getting the picture from these three tubes properly aligned on the screen can be anywhere from slightly to maddeningly difficult to do, and it accounts for many of the awful projection TV images you've no doubt seen in the past.

This process is called setting *convergence*. Some CRT RPTVs have automatic convergence systems that do a great job of setting this up for you. If you have to do it manually, it's quite difficult and best done by a pro. Calling in a pro is not cheap — in fact, it probably takes away any cost advantage you might gain by buying a CRT system instead of a microdisplay.

Tubes have some definite picture advantages over other systems, though:

- **Best *black-level* reproduction:** Blacks are essential when you're watching a scene on TV with dark (or no) lighting and lots of shadows. CRTs are the best of all systems at reproducing black colors on-screen.
- **Excellent color reproduction:** Many microdisplay systems have a particular color that they can't reproduce as well as others — CRT systems don't usually face this issue.
- **No visible pixels or screen door:** Whereas a great many microdisplays have a visible pixel structure, creating a *screen door* effect (meaning you can see the individual dots that make up a picture when you're close to the display), CRTs have none and are therefore smoother and more film-like at close viewing distances.

CRTs also have some negatives:

- **Reduced brightness:** Compared to microdisplays (which use an intensely bright bulb to create light on the screen), CRTs put out fewer lumens.
- **Potentially lower resolution:** Most CRT projectors can't reproduce the full 1,920-x-1,080 resolution of 1080i or 1080p HDTV content. Models with 9-inch CRTs might be able to, but the average CRT RPTV is limited to roughly 1,200–1,300 lines of resolution, rather than the full 1,920. We talk about display resolutions in Chapter 21, in case you're not sure what we're talking about here.
- **Limited tube life span:** Although your CRT projection TV should last years, the tubes do wear out slowly and continuously over time.
- **Susceptible to *burn-in:*** CRTs create a picture when electrons hit a layer of phosphor (which lights up when struck by these electrons). Images that don't move (such as those in video games or even stock tickers) can create a permanent ghost image in these phosphors — effectively ruining your screen.

CRT projection TVs are the most susceptible HDTV for burn-in.

The finest CRT projection TVs (and the most expensive) are some of the really high-end front-projection systems. The key factor in a front-projection system is the size of the CRTs — generally speaking, the bigger the better. Most projectors have 7- or 8-inch CRTs, but a few mega-dollar units (often costing $40,000 or more) have 9-inch CRTs. These are the projectors you find in dedicated home-theater rooms of the truly rich and famous.

Rear-projection CRT-based HDTVs, on the other hand, are a relative bargain. You can easily find a high-quality, HDTV-capable RPTV with a big 50-plus-inch screen for under $1,000 — less than half of the cost of a similar-size microdisplay RPTV.

The biggest disadvantage of CRT-based RPTVs is their sheer size — these things are the behemoths of the HDTV world, particularly when it comes to the depth of the unit (how far it sticks out from the wall behind the TV). Because of the size of the tubes and the complicated mirror systems required to get the image onto the screen, some CRT RPTVs are 3 or more feet deep.

Although we've stated some disadvantages to CRT RPTVs, we're not trying to talk you out of it — many folks *love* their CRT RPTVs and prefer them to microdisplays. If you're considering a CRT rear-projection HDTV, we strongly suggest you find a unit (like several of Hitachi's CRT RPTVs) with an *automatic convergence* system that aligns the picture tubes at a push of a button.

If a CRT projector sounds right up your alley, don't wait too long before you make your purchase. Many manufacturers are abandoning this format for microdisplay projectors.

LCD

If you've ever used a laptop computer, cell phone, PDA, or just about any kind of electronic device with a screen, you've used an *LCD* (or liquid-crystal display). LCDs range in size from tiny, sub-1-inch models to huge 40-plus-inch versions (check out Chapter 23 for examples of these). LCD projection TVs tend to use LCDs from the smaller end of this continuum, often as small as 1 inch or less. Like CRT RPTVs, LCD RPTVs use not one, but three image sources — an individual LCD each for red, green, and blue. Unlike CRTs, however, these LCDs don't require periodic alignment (or *convergence*), which makes owning an LCD RPTV a much easier task for people who don't specialize in TV maintenance.

The small size of these LCDs accounts for what we think is the biggest advantage of the LCD RPTV when compared to a CRT model: LCD models are simply *much* thinner, closer to a flat-panel TV than a traditional RPTV in depth. So they can fit into your tight family room better than an older CRT TV.

LCD projectors have some disadvantages, however:

- **Screen doors:** LCD screens consist of a large number of sharply defined, square-shaped pixels that make up the image. When blown up to big-screen size (generally speaking, 50 inches or greater), these pixels can become visible. You notice this when you feel like you're looking out your home's screen door. Better LCD HDTVs avoid this syndrome, but it can show up on even the best projectors for very large images. Generally speaking, if you're not sitting too close to your HDTV (no closer than two or three times the vertical height of the screen), you shouldn't notice this effect.
- **Less than full 1080p resolution:** As we write, the vast majority of LCD projectors aren't capable of the full 1,920-x-1,080 resolution of the highest HDTV signals. This isn't the end of the world (most LCD projectors have a native resolution of 1,280 x 720 — or 720p), but if you're hoping for the highest resolutions offered by HDTV broadcasts or HD DVD and Blu-ray discs, you won't quite reach it with an LCD projector.
- **Poor black reproduction:** LCDs are a *transmissive* technology — light shines through the LCD. It's hard for LCDs to become totally black (some light leaks though), so dark scenes look more like dark gray than true black. Better LCD HDTVs have better black levels, but none match a CRT projector.
- **Dead pixels:** This is a *huge* deal for some folks — others won't even notice it. Literally millions of pixels make up the three LCDs found in an LCD-projection HDTV. Occasionally, one of these pixels malfunctions or

becomes stuck, resulting in a visible dark or bright spot on your HDTV's screen. The real problem is that many manufacturers won't fix or replace your HDTV if you have only a few of these malfunctioning pixels. So if this sort of thing really drives you crazy, check out the warranty terms before you buy!

- **Lamp death:** The super-bright lamp that powers an LCD projector system has a limited life span — usually after a few thousand hours of use, the bulbs either dim to below usable levels or burn out altogether. Although changing a bulb isn't too hard or too expensive in most cases (usually less than a couple hundred dollars), it is a bit of a pain.

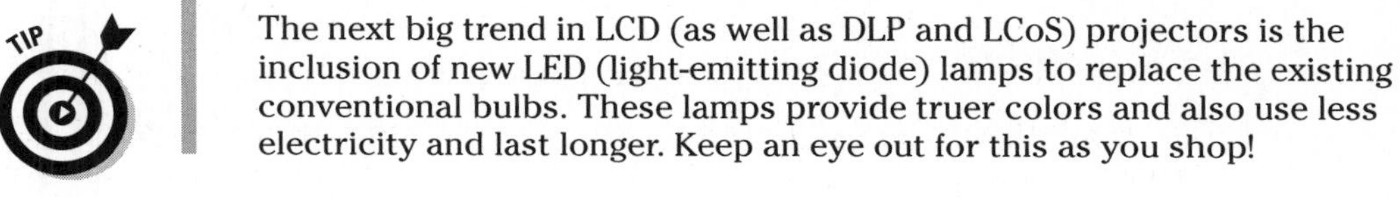

TIP

The next big trend in LCD (as well as DLP and LCoS) projectors is the inclusion of new LED (light-emitting diode) lamps to replace the existing conventional bulbs. These lamps provide truer colors and also use less electricity and last longer. Keep an eye out for this as you shop!

DLP

LCD isn't the only microdisplay technology vying for your projector dollars. Texas Instruments (of calculator and chip fame) has developed a system called *digital light processing* (or DLP), which is perhaps the hottest projector technology available today.

DLP systems are based on *micromirror* technology. A DLP chip (the basis of the projector) is an optical semiconductor with millions of tiny mirrors controlled by the logic portion of the chip. Basically, these little mirrors are individually controlled and tilted to reflect an amount of light corresponding to the picture brightness required for a single pixel of video. The angle of the mirror is changed to move from black (or close to black) — where no light is reflected on-screen — through a whole range of grays right on up to white. Figure 22-4 shows a DLP chip.

Figure 22-4: DLP chips contain millions of tiny mirrors!

With a single DLP chip, color is added with a separate device known as the *color wheel* — most often a set of red, green, and blue filters arranged in a wheel that is located in the path of light reflecting off the mirrors in the DLP chip. The three colors mixed together produce the colors found in your HDTV's source material.

Most people don't notice the single-chip DLP issue known as the *rainbow effect,* but some people with particularly sensitive vision do. The spinning color wheel causes this effect, and can cause a very small percentage of the population to feel dizzy or get a headache while watching DLP — particularly when moving their heads, or during rapidly moving scenes on-screen. Most people *don't* have any problem with this rainbow effect — and most people who own DLP HDTVs love them and never deal with this problem at all!

Some really expensive DLP projectors (like those used in movie theaters) use three DLP chips (one each for red, green, and blue) instead of a color wheel. This system reproduces an even greater number of colors and tends to provide a smoother image on-screen and eliminates all possibility of the rainbow effect. You don't *need* three chips to eliminate the rainbow effect. Some manufacturers have introduced a six- or seven-element color wheel to overcome the defect. Mitsubishi also added a magenta to the red, green, and blue seven-segment wheel, which results in much-improved color.

As we write, several LED light source DLP rear-projection systems have been announced and are hitting the market. These HDTVs use a single DLP chip but replace the traditional bulb light source and the spinning color wheel with a lighting system that uses three LEDs (red, green, and blue) to create the picture. These systems have three big advantages:

- You lose the rainbow effect because there's no more color wheel.
- You get better color reproduction because the LEDs are simply better at creating the full range of colors than is a "white" light bulb and color wheel combination.
- You no longer have to worry about replacing a bulb (which can cost a couple hundred dollars) because the LEDs should last for up to 60,000 hours of viewing.

DLP-projection TVs are about the hottest thing on the market today. They're very thin (some less than 6 or 7 inches — nearly in plasma territory), and produce a bright, beautiful picture, with black reproduction that's better than LCD and excellent color reproduction.

Wobbling away!

There are two ways to get 1080p resolution out of the DLP chip design. The first is to create a chip that has 1,920 x 1,080 micromirrors on the chip, which is a pretty straightforward way of doing things! The second is to have slightly fewer individual micromirror "pixels" on the chip but to use them in a special way to make the individual mirrors reproduce more than one pixel on your screen.

This second approach, called *wobulation,* was developed by the folks at Hewlett-Packard, which, by the way, sells its line of DLP HDTVs along with printers, servers, laptops, and every other imaginable bit of computer-related electronics (`www.hp.com`). At a high level, wobulation is sort of like interlaced video because it relies upon the persistence of vision, meaning you see something longer than it's actually in front of you, especially when images are changing very rapidly, like every 1⁄60 of a second. A DLP chip with wobulation uses another mirror that moves even more quickly (once every 1⁄120 of a second, or faster) to re-aim the pixels from individual mirrors on the DLP chipset. The effect to your eye is 1080p resolution out of a chipset with something less than 1,920 x 1,080 micromirrors (TI won't say how many it has).

Many of the 1080p rear-projection DLP HDTVs on the market use this technique to gain their high resolution. It's very effective, and looks great on the screen. However, because Texas Instruments also has a DLP chip available with "true" 1080p resolution (no wobulation), we expect that over time you'll see more and more models switching to this newer version of the DLP chip. The wobulation-based chips are cheaper, however, and we wouldn't steer you away from buying an HDTV based on these chips if it fits your overall needs.

When you're choosing a DLP-based projection system, be sure to read the fine print. Not all DLP systems are HDTVs — some inexpensive projectors (mainly front-projection systems) use older DLP chips that don't reach HDTV resolutions. The latest DLP chips support the full resolution of 1080p (1,920 x 1,080). Older and less expensive versions provide 1,280-x-720 resolution (which perfectly matches 720p resolution requirements). Check out the sidebar titled "Wobbling away!" for more on the 1080p chips.

LCoS

The newest projector technology on the block is called *liquid crystal on silicon* (or LCoS). Although only a few LCoS HDTVs are available on the market today, they're coming from some very major companies (particularly Sony with its SXRD system), and they're reviewed as some of the finest HDTVs on the market in terms of picture quality.

The LCoS (which combines liquid-crystal technology, like an LCD, and a *reflective* instead of *transmissive* system) is designed to be a cheaper way of getting to really high-definition HDTVs. Whereas most LCD and DLP systems are limited to 720p resolutions (they convert 1080i signals to 720p for playback on-screen), some LCoS systems can display the full resolution of 1080i. That means that an LCoS system can display 1,920 x 1,080 pixels — something no other consumer HDTV system can do.

The two main manufacturers of LCoS HDTVs today are Sony and JVC. Sony's SXRD line includes both front- and rear-projection 1080p HDTVs, whereas JVC's HD-ILA models currently offer 720p resolutions. We expect that JVC will soon offer 1080p HD-ILA HDTVs and that other manufacturers will join in the LCoS game as well.

Screens

Whereas RPTVs have a built-in screen, front-projection TVs require a separate screen system for displaying the image. You can use a wide range of materials for a screen — some people just use a white-painted wall or a tautly hung white bedsheet, for example — but most people choose a commercially built TV screen.

You can find screens that are *fixed* (permanently mounted) — these are usually the best value and the best performers. Unless you have a dedicated home-theater room for your HDTV projector, you might want a *retractable* screen (like the ones in your fifth-grade classroom, only these are often electrically powered) or *tripod* screen (that you can fold up and put in the closet when not using).

Even more important than the form are the technical characteristics of the screen. The big three are the following:

- **Gain:** *Gain* is a measure of how reflective the screen is — how much of the projector's light gets bounced back to your eyeballs. There's a standard industry reference for gain, and systems that have exactly as much gain as that reference are rated at a gain of 1. More-reflective (high-gain) screens are rated greater than 1 (say 1.2), and less-reflective (low-gain) screens are rated below 1 (many are rated at 0.8). If you have a CRT projector, get a high-gain screen rated between 1 and 1.3, though you can go higher if needed. Microdisplay projectors can use a low-gain screen (0.8 or higher) due to their brightness.

- **Viewing angle:** Most display systems have a limited angle (from perpendicular) in which they look best. If you sit outside that angle, the picture becomes very dim. Viewing angle is inversely proportional to gain. In other words, higher-gain screens have smaller viewing angles. For this reason, you're best off choosing the lowest-gain screen that works with your projector. Microdisplay projectors have light to spare, so a low-gain screen is worthwhile to gain (pardon the pun) a bigger viewing angle. Viewing angles are usually listed in a number of degrees (say, 90). Your viewing angle is half this amount (45, in this case) on either side of perpendicular.
- **Aspect ratio:** Screens are available in either the 16:9 widescreen or 4:3 aspect ratios. You should choose the same aspect ratio as that of your projector — which, with any HDTV, is 16:9. You might want to buy *masks* to place on the sides of your screen when you're watching 4:3 material (such as older, standard-def TV shows). These masks — similar to mattes for a picture frame — can help avoid any leakage of light beyond the edges of your picture.

Chapter 23

Thin Is In

In This Chapter

- Getting into flat panels
- Understanding the pros and cons
- Looking into LCDs
- Picking a plasma screen

Flat-panel TV technology — super-thin HDTVs that you can hang on the wall like a painting — have really taken a prime place in the popular *zeitgeist.* A big-screen, flat-panel plasma or LCD TV has become *the* status symbol of the '00s, not just in the living room, but in hotel lobbies, retail stores, and even in the backs of cars.

There's a good reason for this mania: Flat panels provide a large viewing area with almost no intrusion into your HDTV viewing room. If that was all they did, they'd be pretty cool, but they can offer a very high-quality HDTV picture, as well.

In this chapter, we talk about flat panels and specifically about plasma HDTVs and LCD flat-panel HDTVs. We also discuss the pros and cons of each type and compare them with other more traditional HDTVs.

Many flat-panel displays ship as monitors only — with neither an HDTV nor an NTSC tuner included as part of the display. Remember that you need some sort of external HDTV and NTSC tuner (or a cable set-top box or satellite receiver) to view TV on these flat-panel displays.

Loving Your LCD

We're willing to bet that you're already familiar with the LCD display. If you have a flat-panel display for your PC, laptop PC, PDA, cell phone, GameBoy, or just about any digital device with a display, you have an LCD. LCDs have been around for decades, mainly in lower-resolution formats and smaller sizes (such as phone screens), but they're getting larger all the time — and growing sharper in resolution.

We're talking about *direct-view* LCDs here — in other words, LCD HDTVs where you look at the LCD display itself. In Chapter 22, we talk about projectors with LCD microdisplays — teeny tiny LCDs that are used to project a bigger image on a screen. Figure 23-1 shows an LCD HDTV.

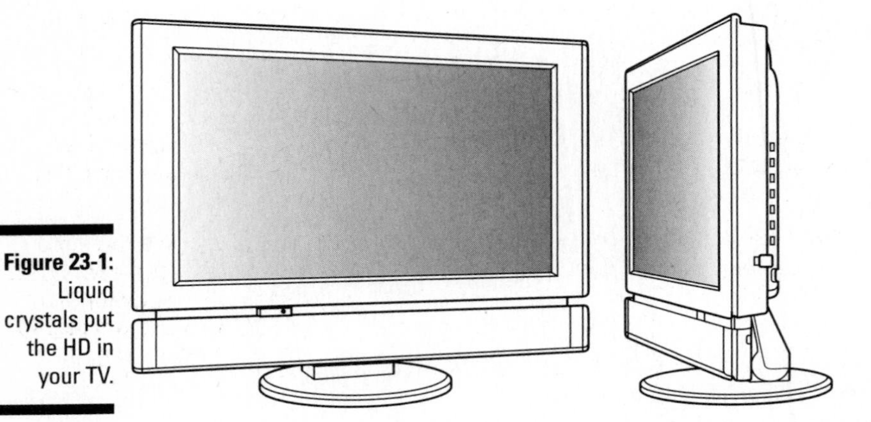

Figure 23-1: Liquid crystals put the HD in your TV.

Although the LCD TVs we discuss in this chapter are a lot bigger than those inside of a microdisplay projector and bigger than those in your laptop or desktop computer, they're still typically not *really big.* The process of "growing" LCD crystals is a manufacturing challenge, and bigger LCDs are harder to make because they have lower *yields,* or percentage of LCD crystals that can be used for HDTVs. Because of this, most LCD HDTVs on the market are smaller than 42 inches — in other words, smaller than projectors or most plasmas (discussed in the second half of this chapter).

A few large-screen LCD HDTVs are on the market these days, such as Samsung's 52-inch model, and we've seen announcements of a number of 50-inch and larger LCD HDTVs. Typically, however, LCD gives way to plasma when you hit the 50-inch mark for thin TV displays.

Not only are LCD HDTVs growing to huge dimensions, they are also becoming huge sellers. A few reasons why include the following:

- **High resolution:** Some LCD displays are 1080p HDTVs, capable of reproducing every single pixel of 1080i/1080p HDTV programs, which means you get 1,920 x 1,080 pixels. In fact, although a few rear-projection and plasma models are also capable of this high resolution, 1080p is most commonly found in the LCD world. Keep in mind, however, that not all (not even *most*) LCD displays are 1080p. Many are 720p displays with resolutions of 1,366 x 768 (slightly higher than 720p, actually, and the same as many plasma displays).

- **Excellent color:** LCDs can display millions of colors, and they do so accurately (meaning the color coming off the screen is faithful to the color in your broadcast or recording). Not all flat-panel TVs or even HDTVs in general can duplicate this color accuracy.
- **PC-monitor-capable:** You can also use many LCD HDTVs as big (huge!) PC monitors. This trick is especially cool if you have one of those neat Media Center PCs we discuss in Chapter 16.
- **No burn-in:** HDTVs that rely upon *phosphors,* such as CRTs and plasmas, can under certain circumstances experience *burn-in,* where ghost images are permanently burned into the screen. LCDs are immune to this phenomenon — so feel free to play video games, watch the CNBC stock ticker, and so on with no fear.
- **Inherently progressive:** Unlike tube (CRT) TVs, LCDs don't rely on a scanning "gun" and *interlaced* scanning (see Chapter 21). Instead, LCDs use millions of tiny transistors that can be individually controlled by the "brains" inside the display. This means that LCDs can easily handle progressive-scan sources, such as progressive-scan DVD and HDTV.

Making the plasma versus LCD decision

Plasma and LCD flat-panel displays both have their proponents. Those in the LCD camp point to the large number of 1080p displays, the excellent color, and also the fact that LCD TVs tend to be a bit more energy efficient. (Fight global warming while still enjoying your HDTV!) And LCDs are immune to screen burn-in and may actually last a bit longer than plasmas.

On the plasma side of things, proponents first point out that when you want to get a really big flat-panel display (60 inches or more), plasma is the only game in town. And for the smaller, but still-big sizes (like 50-inch displays), plasmas are simply cheaper than the same-sized LCD HDTV. (Though the price difference has been dropping a lot over the past few years, plasmas are still cheaper on a per-inch basis.) And plasmas have a wider viewing angle and a picture that many feel is superior.

Where do we come down on this debate? Well, we think it's really an HDTV-by-HDTV comparison. We've seen stunning pictures from both types, and every year the picture gets better and the price comes down. If you're looking for a big screen (for example, if you're sitting 12 feet or more from the display), a plasma is still probably your best bet, but over time that difference will probably go away. For now, we say look at both with an open mind and find the HDTV that looks best to you while fitting your budget.

In the long term, new technologies are arising, such as *SEDs,* or surface-conduction electron-emitter displays, which use less power, provide CRT-like blacks, and still hang on the wall like a plasma or LCD. These new technologies might overcome both types of displays in the marketplace and on your living room wall. For now, LCD and plasma are both great!

We don't include the super-skinny-hang-it-on-the-wall attribute here, because we think that's obvious. We can't resist mentioning it again anyway, just because it's so cool.

Besides the size limitation we discuss earlier in this section, you might want to consider a few other issues before you choose an LCD HDTV:

- **Expensive for their size:** LCD HDTVs are great, but they're not cheap. Of course, all flat-panel TVs are relatively pricey, but LCD HDTVs typically cost more, per inch, than do plasmas. In the summer of 2006, you can expect to pay about $1,800 for an HDTV-ready, 37-inch LCD display. For the same amount, you can easily pick up a 42-inch high-definition plasma, or perhaps even a 50-inch model.
- **Poor reproduction of blacks:** Black images are among the hardest for most TVs to reproduce (CRT TVs are best at this). LCD TVs tend to produce grays, not blacks. LCD HDTVs lose out, in this regard, to both CRT and plasma TVs.
- **Limited viewing angle:** LCDs typically have a poor viewing angle — the angle you can sit away from perpendicular and still see a clear image on-screen. Manufacturers have been working diligently to improve this characteristic (with some success). Check the specs before you buy — most LCD HDTVs have viewing angles listed in their specifications.

  Viewing angles aren't only horizontal (left to right); they're also vertical (top to bottom). If you're hanging an LCD HDTV on the wall, the vertical viewing angle might be more important to you than the horizontal. Most people pay attention only to the horizontal viewing angle — don't be like most people.
- **Slow pixel response time:** Another area that LCD HDTV makers are working overtime to improve is the *pixel response time* of their TVs. Basically, the individual pixels within an LCD HDTV take a slight amount of time to change color and intensity. For really fast-moving video content (particularly in a 720p picture, where every pixel can change as many as 60 times per second), an LCD TV can end up with some *artifacts* (visible flaws) where the picture from a previous frame is still slightly visible on-screen as the new one is being drawn. Typically, artifacts aren't a huge and noticeable deal, but it's not beyond the realm of possibility that you might notice it.

  Look for an LCD with a *low* pixel response time. The best models today have pixel response times as low as 8 microseconds, which pretty much guarantees you a picture totally devoid of these artifacts.
- **Limited brightness:** The LCD is a *transmissive* system — light is shined through the liquid crystals. Some of the light gets absorbed or reflected back, away from the viewer. This means that LCD displays aren't as bright as CRT, plasma, and even some projection TVs (DLP, for example). This could be a factor in a brightly lit room.

A few of the fanciest LCD displays now use LED (light-emitting diode) backlights instead of the more traditional fluorescent bulbs. These models are both brighter and have better color reproduction. Also an LED backlight will last longer than a traditional (bulb-based) backlight, increasing the already long life span of your TV.

Everyone's Crazy about Plasma

The really hot spot in the HDTV technology market is the plasma TV. Plasma TVs combine a thin, compact chassis with a truly large (even huge) screen size and then add beautiful high-definition pictures to the mix. For many potential HDTV buyers, plasmas really fit the bill.

A plasma screen contains literally millions of gas-filled cells, each one acting as a single image pixel, trapped between two pieces of glass. An electrical grid zaps these cells and causes the gases to ionize, and ionized gas is plasma — hence the name. The ionized gases, in turn, cause a layer of phosphor on the viewer's side layer of glass to light up (just as the electron gun in a CRT causes the phosphor to light up on the front of the tube).

Despite their compact dimensions (in the depth direction at least; many plasmas are only about 4 inches deep), plasma HDTVs are available in 42-, 50-, and even 60-plus-inch sizes. Imagine a 4- or 5-inch-deep HDTV that spans 5 feet diagonally, and you can see the instant appeal of plasma.

Other benefits of plasma displays include

- **Excellent brightness:** Plasma HDTVs are second only to CRT direct-view TVs (discussed in Chapter 24) in terms of picture brightness. Plasmas don't rely on a light bulb shining through or reflecting off of something as an LCD or DLP system does. In some ways, plasma brightness is even better than the brightness you get with a CRT because the picture is uncannily evenly bright across the entire screen. In a CRT, on the other hand, there always is some slight (or not-so-slight) difference in brightness as the electron beam reaches different parts of the screen.
- **High resolution:** HDTV plasma TVs can often reach higher horizontal resolutions that CRT-based direct-view sets just can't match. The finest plasma TVs have such high resolutions and such smooth images that they look like nothing more than beautiful film images. Some manufacturers (like Pioneer, with its PRO-FHD1 50-inch plasma HDTV monitor) can reach the full resolution of 1080i/p HDTV sources with their panels. Most HDTV plasmas have a 720p resolution (or just above that) of 1,366 x 768.

Just because it's plasma and it costs $1,500 or more doesn't make it an HDTV. We don't know of any sub-$1,500 plasma that *isn't* an EDTV (enhanced-definition TV) *rather than an HDTV.* This will change eventually, but for now, "cheaper" plasmas are EDTVs.

- **Progressive by nature:** Like LCD displays, plasma systems don't use a scanning electron beam to create a picture. Instead, all the pixels on the screen are lit up simultaneously. Progressive HDTV sources (such as 720p broadcasts or Blu-ray and HD DVD discs) and non-HDTV sources (such as progressive-scan DVD players) are displayed to full advantage on a plasma HDTV.
- **A wide viewing angle:** Unlike LCDs, which often have problems in this regard, plasma displays have a good picture even when you're sitting *off axis,* or not perpendicular to the screen surface. This is a huge benefit for smaller rooms, where viewers might sit relatively far off to the sides of the screen, at wider angles.

Plasma's not perfect, of course:

- **Susceptible to burn-in:** Any system that uses a phosphor screen to display video can fall victim to the phosphor burn-in. If the Xbox is a primary HDTV source in your home, consider something besides a plasma — maybe an LCD HDTV or a rear-projection microdisplay HDTV.

  You can minimize burn-in on any display by calibrating the set properly and reducing the brightness from its (usually too-high) factory setting. And don't worry *too much* about burn-in, as most plasmas today have good internal systems to reduce the incidence of it. But don't ignore it entirely, because burn-in can still happen.
- **In most cases, less than the highest resolution:** Most plasma HDTVs are *720p* sets. (Actually, the resolution of a typical 50-inch plasma is 1,366 x 768 pixels, just a hair above 720p.) Only a few plasma HDTVs on the market as we write are capable of the full 1080p resolution of 1,920 x 1,080 pixels. For folks looking for this highest possible resolution, a few high-end 1080p plasma displays have become available on the market in mid-2006 — at a significantly higher price than the more common 720p models. The Pioneer PRO-FHD1, with a native resolution of 1,920 x 1,080, runs you $10,000 — about four to five times the price of many 720p plasmas of the same 50-inch size.
- **Shorter life span:** Another phenomenon of any phosphor-based display system is that eventually the phosphors lose their brightness. This process is subtle and slow, but it inevitably happens. If you've saved up to buy an HDTV to last you a lifetime, well, don't get a plasma unless your personal actuary tells you that you're close to the end of your rope.

  Before you buy, check the manufacturer's specifications on *hours to half brightness* (the point at which the display is only half as bright as it was when new). For example, if this specification is 20,000 hours, and you watch the set for 6 hours a day, it will effectively wear out in about 9 years. If you have kids, keep in mind that 6 hours a day is (comparatively) *not* a lot of time for the TV to be on every day.
- **Less-than-perfect color reproduction:** Although plasma displays are capable of producing a breathtaking array of colors, all the sets built to

date have had an unfortunate tendency to make red colors look more orange than true red. If you're a huge fan of slasher horror flicks, this might take away some of your fun!

- **Poor reproduction of black:** Although plasma displays are nearly equal to CRT sets in terms of absolute brightness, they fall short in the realm of reproducing black images. Most plasmas do a slightly better job than LCD TVs at black reproduction, but they fall short of CRTs and some projection systems (such as DLP).

Going commercial

Even though prices have plummeted in the past few years, plasma HDTVs are still relatively expensive when compared with most other types of HDTVs. In fact, inch per inch, only LCD HDTVs cost as much or more than plasmas. So many folks who are taken with the great picture and sleek profile of plasmas look for a way to get one without breaking the bank.

One interesting approach is to skip over the "consumer" plasmas found in traditional consumer electronics retailers and to instead look at a *commercial plasma.* These are the plasma displays you see in hotel lobbies, airport flight status displays, and by the hundreds on trade show floors. (It seems like *every* trade show we attend these days has at least one plasma per company booth.)

There's often very little substantive difference between a consumer and a commercial plasma display, at least in terms of the features that influence your picture quality. In fact, under their surfaces, most commercial plasmas are identical to their consumer cousins, with only styling and feature differences.

So what are those feature differences?

- **Most commercial plasmas don't have any type of TV tuner (not even an NTSC tuner for standard definition broadcasts).** You need to use an external TV tuner, cable box, or satellite box to receive standard- or high-definition TV broadcasts on these models.
- **The manufacturer assumes you'll mount the commercial plasmas on a wall, a ceiling, or even on an airport luggage carousel.** There likely won't be any kind of a basic stand in the box for setting up your plasma on a piece of furniture.
- **Many commercial plasmas are designed to connect to a wider variety of source devices than their consumer cousins.** For example, Panasonic — which is widely known for its commercial plasmas — offers a wide variety of boards for its plasma's multifunction slots. This gives you a great amount of flexibility in what you use to "feed" your plasma, from RGB connectors for home-theater/Media Center PC connections to HDMI and way beyond! Panasonic even offers an built-in PC that slots into one of these modules, which might not be too applicable for your home theater, but we still think it's pretty darned cool.

The best part about these commercial plasmas? They're typically 10 or 20 percent cheaper than the consumer variants of the same plasma panels. If you don't mind a bit of extra work installing input boards (it's a simple process, much like adding memory or other components to a PC), and you don't need a built-in tuner, you might be very happy getting all industrial and going commercial. Check out online vendors like Visual Apex (`www.visualapex.com`) for great deals on these plasmas.

Taking a plasma uphill

Plasma TVs rely upon a thin layer of gas (which gets ionized and eventually turns into the picture you see). If you think back to your chemistry classes in high school and college, you might start remembering . . . well, let's not go there. You might also, however, recall concepts like the *ideal gas law* (PV=nRT), which basically says that (all else being equal) the volume of a gas is inversely related to its pressure. Which is a nice way of saying that if the atmospheric pressure goes down (as it does at high elevations), a gas expands.

As some folks living in high-altitude cites have discovered, this expanding gas can cause the glass screen in a plasma to bow outward and potentially vibrate and make an annoying noise. If you live some place high (say a mile above sea level or higher), check the manufacturer's specs before buying a plasma set. In response to this, manufacturers have begun to make specific design changes in reaction to this situation and have begun selling plasmas that won't get altitude sickness.

The bottom line here is that you might hear stories about plasmas having issues at higher elevations. We've even heard people recommend against plasma TVs because "you'll need to change the gas." Whatever altitude or gas issues plasmas have had in the early days have been effectively solved with the latest models.

Every type of HDTV has its pros and cons. We're certainly not trying to talk you out of going the plasma route (if your budget can handle it); just making sure that you're informed. We doubt that you'll find anything to dislike about the picture and performance of the best HDTV plasma sets — with the possible exception of the price tag!

Chapter 24

Good Ol' Tubes

In This Chapter

- Understanding the CRT
- Figuring out the pros and cons
- Choosing an aspect ratio
- Doing the squeeze
- Going flat

You've seen and watched *CRT* (cathode-ray-tube) TVs — they're the traditional tube TVs that are in hundreds of millions of homes around the world. CRT is an old-school technology that's been used for TVs for more than 60 years, but that doesn't mean that the old box doesn't still have some life left in it.

In fact, CRTs in many ways offer the best pictures available in the HDTV world. They aren't the biggest, they aren't always the highest in absolute resolution, and they're certainly not the slimmest or thinnest HDTVs, but just about nothing can beat a CRT HDTV when it comes to brightness, reproduction of colors, and the ability to show darkly lit scenes (black reproduction).

In this chapter, we start with an overview of how CRT TVs work and what they are. We follow with the pros and cons of the CRT as an HDTV. We finish with aspect ratios. CRTs are the one kind of HDTV screen where you often find 4:3 aspect ratios. We tell you why that is and what you should consider if you're buying a CRT HDTV.

Many (most, in fact) big name-brand manufacturers of CRT HDTVs have begun to shut down the production of these models and are instead expanding the production of LCD and plasma flat-panel displays (see Chapter 23) or microdisplay projection TV systems (see Chapter 22). For example, Sony — probably most famous in recent years for its *Trinitron* CRT TVs — has begun shutting down its gigantic tube factory near Pat's home in San Diego. The reason has less to do with the quality of a CRT picture, which can be excellent, than simple market trends — people are willing to pay more for a slim and sexy flat panel, and the profits simply aren't there for CRTs. What we're getting at here is that every month that goes by brings fewer and fewer choices in the CRT HDTV world. So if you're going to buy, don't wait forever (or don't be surprised if you have very few options).

All about CRTs

The CRT consists of three major elements (sure, it has a bunch of other small and significant parts, but the ones listed here are the key parts):

- **The vacuum tube:** This is why people call a TV "the tube." The major structure inside a CRT HDTV is a large glass vacuum tube. You might also hear these referred to as *direct-view* TVs — direct view as opposed to projection, meaning you watch the image directly. This tube is what makes a CRT HDTV simultaneously heavy and fragile.
- **Phosphor coating:** On the inside of this vacuum tube, on the opposite side of the screen from where you view the TV, is a coating of *phosphors.* These molecules emit light when they're hit with electrons, and this light is the picture you see.
- **An electron gun:** This device at the back (or *neck*) of the tube is the source of the electrons that pummel the phosphors on the front (screen) of the tube. The electron gun controls this beam of electrons by aiming them (electromagnetically) into a series of *scan lines* that run across the screen.

A color CRT actually has three electron guns — one each for red, green, and blue. This combination of colors (often called *RGB*) produces any color on the screen.

CRT Pros and Cons

Because CRT TVs are a mature technology, lots of folks don't give them a second (or first!) thought when they're shopping for a new HDTV. We think it makes sense to at least consider a CRT HDTV. Some reasons include the following:

- **Best at black:** When it comes to producing video that accurately reproduces dark scenes (such as the shadows in a horror movie when the bad guy with a chainsaw or axe is sneaking up on the nubile teenage girls — Danny's favorite scenes!), CRTs rule. Most other HDTVs can only produce grays that approximate black — CRTs can give you something like inky black!

- **Works well in a bright room:** Cathode-ray tubes really blast out the light. Some of them are so bright they could almost replace the fresnel lens in a lighthouse (well, not really). This means that CRT HDTVs are perhaps the most suitable of all types for use in a brightly lit room. So if you're mainly watching sports in a family room with floor-to-ceiling windows, consider CRT.
- **Easy on the wallet:** Because the factories and machines and technologies behind CRT are old (though constantly refined), CRTs are rather inexpensive to build. HDTV CRT TVs are, of course, more expensive than their analog bretheren, but still relatively cheap. You can find a 32-inch CRT HDTV for well under $1,000 — under $700 for smaller, bedroom-sized models. You're simply not going to find a plasma, LCD, or projector HDTV for that little cash (unless you're talking about a really, really tiny LCD TV).

Nothing's perfect, and CRTs have some downsides, too. These are some that get us in a lather:

- **Size does matter:** That is to say, if you want a really big-screen TV, CRT might not be for you. It's a lot easier, as an engineering and manufacturing challenge, to create a big-screen TV by using a projector system, or even a plasma or LCD system, than it is using a CRT. Huge vacuum tubes aren't easy to design, and they're even harder to build. The biggest widescreen CRT HDTVs on the market today are 34 inches. A few traditional 4:3 aspect ratio 36-inch models are still available, but many of them aren't HDTVs. If you're going to be sitting more than 10 feet from your HDTV, consider something besides a CRT.

  Another problem with big CRT TVs is making sure that the electron beam is accurately aimed at the right places on-screen. The bigger the screen is, the harder it is to keep the picture consistent across the entire screen. This is just a matter of geometry — the edges of the screen are farther away and at a more obtuse angle from the gun(s) than is the center of the screen.
- **Portly TVs:** We're sure you've seen and probably lifted enough CRT TVs (be they HDTVs or just analog sets) to know this problem by heart. CRT TVs are really heavy, especially when their size moves beyond the 30+ inch mark. We mean REALLY heavy — back-breakingly, "Hey Ma, where's my hernia belt" heavy. They're also fat because the geometries required by the tube take up a lot of depth. You'll never hang a CRT on the wall (at least not without some huge bridge girder-like stand), and it'll never be slim and sexy like a plasma or microdisplay projector system. So to give you an example, the Sony's KD-34XBR970, a 34-inch widescreen model, is 25.75 inches in height, 39.25 inches in width, 25.25 inches in depth, and weighs over 190 lbs . . . and that's typical.

  The reason big CRT HDTVs are so heavy has to do with the vacuum tube. To get the tube so big without having it break under its own weight, the glass must be extra thick, and extra thick in this case means extra heavy.

Some manufacturers, notably Samsung (`www.samsung.com`), have created CRT HDTVs that are slimmer. Samsung claims that the SlimFit models, for example, are 25 percent thinner than the average CRT TV of the same size.

- **Lower resolution:** Although CRT HDTVs can have extremely high resolutions, most don't reach the levels of resolution found in the best LCD, DLP, plasma, or LCOS systems, or even to the resolution that the most expensive CRT-based front-projection TVs reach. Where direct-view CRTs fall a bit short is in the horizontal resolution — the number of pixels running across the screen from left to right. For example, the 1080i format can use up to 1,920 pixels in this direction; very few CRT systems can focus in on that large a number of distinct pixels horizontally. Most max out in the range of 800 lines of horizontal resolution, and a few models (like Sony's "Super Fine Pitch" HDTVs) can get over 1,000 lines, but these models are beginning to be discontinued and are hard to find.

If you can spend some extra money, you might consider checking out a CRT-based projection TV, discussed in Chapter 22. These systems cost more and take some extra work in terms of setup and ongoing maintenance, but they provide the picture benefits of a CRT HDTV with the big-screen advantages of a projection system.

Unlike basically all of the other types of HDTVs we discuss in this book, CRT-based HDTVs *don't* have a fixed display resolution. They do, however, have optimal or preferred display resolutions — usually one standard def and one high def. Most CRT HDTVs display images either as 480p (standard-definition) or 1080i (HDTV!) — or in some cases 540p, which is half the resolution of 1080i, but in progressive-scan format.

Most CRT HDTVs accept any sort of HDTV signal (such as 720p) and convert it to 480p or 1080i — that is to say, they have an internal *scaler* to change the *input resolution* into a preferred *display resolution,* but other sets require you to send a 1080i signal into the back of your HDTV. Make sure you know what your CRT HDTV requires, and make sure that your HDTV tuner, set-top box, or satellite receiver can be configured to output 1080i — even when the program itself is broadcast in 720p. If your CRT HDTV has a built-in HDTV tuner, you can ignore this advice!

Navigating the CRT Jungle

If you decide to go with a CRT HDTV, you have to decide among a couple of features and attributes.

Dealing with aspect ratios

Most other types of HDTVs are available in only one aspect ratio — the widescreen 16:9 aspect ratio. Direct-view CRTs differ in this regard because HDTV-capable TVs are widely available in both 4:3 and 16:9 aspect ratios.

Which to choose

In the past (and we're talking about just a year or two ago), we often recommended that CRT HDTV buyers spend their money on 4:3 HDTVs with anamorphic squeeze functionality built in (see the next section, "Doing the squeeze"). But that was then, this is now, and we're changing with the changing times.

See, one of the big reasons we recommended 4:3 CRT HDTVs was because there simply weren't a lot of choices. But CRT manufacturers have tooled up what's left of their factory floors to produce more 16:9 tubes, and now most CRT HDTV manufacturers are offering 30-inch and 34-inch widescreen CRT HDTVs.

The other thing that has helped change our minds is the fact that simply more widescreen content is available these days (and more arrives almost daily, it seems).

What we're getting at here is that we now generally recommend a 16:9 CRT HDTV over a 4:3 model for most people. We still like 4:3 models for a couple of reasons:

- They're cheaper, which is always a good reason if your budget (like most people's) is a bit tight.
- For the same size, you get more screen area on a 4:3 TV. This is just a matter of math: height × width = area. A 32-inch 4:3 set (which costs less) has 1 square inch more picture area than a *34-inch* 16:9 model, and it probably costs 20 percent less to boot!

We recommend you get a 4:3 HDTV *only* if it has the anamorphic squeeze function (see the next section). Fortunately, just about every model we know of today has this function.

Many manufacturers have phased out their 4:3 tube HDTVs but still sell models with 4:3 aspect ratios, component-video inputs, progressive scan, and widescreen anamorphic squeeze modes. Don't mistake these for HDTVs, though they can be very high-quality and inexpensive TVs for watching DVDs.

Table 24-1 shows the typical sizes you find for tube HDTVs, and their aspect ratios. Note that we've not seen *every* tube HDTV made in the world, but these sizes are common to all the major manufacturers.

Table 24-1 Tube HDTV Sizes and Aspect Ratios

TV Size (Diagonal Screen Measurement)	*Aspect Ratio*
20"	4:3
26"	16:9
27"	4:3
30"	16:9
32"	4:3
34"	16:9

These size to aspect ratio correlations don't necessarily apply to other types of HDTVs (like LCD flat panels, where 32-inch widescreen 16:9 models are common).

Doing the squeeze

If you choose a 4:3 television, you really should make sure that you choose one with a feature called *anamorphic squeeze* (also called *raster squeeze* or 16:9 mode). This function allows you to watch *full-resolution* widescreen content (HDTV or DVD-based) on a 4:3 HDTV screen.

Here's what the squeeze function does: While you're watching widescreen HDTV or DVD content, your HDTV re-aims its electron gun so that the gun beams scan lines onto the screen phosphors in a 16:9 window on the screen. This re-aimed picture looks, at first glance, exactly like the traditional *letterboxed* picture that old-fashioned analog TVs use to display widescreen content — with the black bars on the bottom and top of the picture. The difference is that with a letterboxed picture, your TV actually wastes some of its scan lines to draw those black bars on-screen, so your actual picture resolution is decreased. When an HDTV uses the squeeze function, all scan lines (usually 1080 of them) are used for the picture. The black bars are simply unlit parts of the screen. The result is a clearer, sharper picture.

Some displays have an automatic system for detecting 16:9 sources and turning on this mode, and others have a button on the remote to activate it. We like the ones with the automatic mode, of course. We're lazy.

This concept wouldn't work with most other displays — which are *fixed-pixel* displays, with a set number of pixels in both directions, horizontally and vertically. The pixels and scan lines on a CRT, on the other hand, aren't determined by the shape of the screen but are instead controlled by the electron gun, which itself can be controlled and can change how it aims its beam. In a fixed-pixel display, you can change between 16:9 and 4:3, but not without leaving some pixels unused (in other words, leaving some of your resolution on the table).

If your HDTV has this mode, make sure you've set your DVD player to "think" it's connected to a 16:9 TV. Otherwise you don't get anything from your display's anamorphic squeeze function because the DVD player plays widescreen movies in letterbox mode, not anamorphic. (If all this anamorphic talk confuses you, check out Chapter 11, where we discuss it in more detail.)

Getting flat

CRT HDTVs will never be flat-*panel* TVs like plasmas and LCDs — there's simply too much going on behind the scenes (or in the back of the tube) to make a CRT that hangs on the wall and leaves you room for a nice game of Nerf hoops in the family room. But CRT TVs can be flat-*screen* TVs.

That is to say, the very front surface of the TV tube — the screen itself — can be manufactured so that it is truly flat.

Older CRTs (and, in fact, most CRTs still sold today) had curved screen surfaces — which made it a bit easier to design the electron gun and such. A flat-screen surface, however, has the distinct advantage of not picking up nearly as many reflections or as much glare from light sources in your viewing room.

A flat screen doesn't make your picture better per se — it just makes it easier for you to enjoy that picture because you aren't squinting through some awful glare to see what's on-screen.

Part VII
The Part of Tens

In this part . . .

Top ten lists! Don't believe David Letterman when he says he invented these — we really did. Okay, we didn't, but neither did Letterman, so there.

In this section, we suggest ten great places to go looking for your HDTV, from highly rated online superstores, to really informative and consultative specialty sites, to your local electronics store. There are pros and cons to buying at various places, so read this before you buy anything.

When it comes to really getting the most out of your HDTV, our chapter on cool HDTV accessories is going to really hurt your pocketbook. From things that make you go bump in the night to 20-x-10-foot screens for your backyard, you're going to want to get all of these items.

When you're done with this part, you're truly ready for the HDTV World. If there were an HDTV Passport, we'd issue one to you!

Chapter 25

More than Ten Places to Buy an HDTV

In This Chapter

- Choosing the best, not just the cheapest, place to buy an HDTV
- Watching your warranty coverage when ordering online
- Finding great deals in bundles

Buying an HDTV is like buying any other major piece of new electronic equipment: It can be extremely nerve-racking to try to balance the desire for the best possible features and systems with the innate fear of finding out too late that you've paid too much or bought something that is instantly obsolete.

Complicating this situation is the wide variety of sources and prices for HDTV systems, making simple comparisons difficult to find. HDTV systems can come packaged with other services, such as cable or satellite service. HDTVs can be sold along with a computer system or sold separately on auction sites. They're easily accessible through online buying services as well as at the traditional electronics stores. HDTV is everywhere.

Each approach has advantages and disadvantages — no single answer fits every consumer, any more than one HDTV fits every living room. The good news is that with so many options, one will fit your needs, regardless of where you live.

Speaking of geography, note that our top-ten list doesn't include regional electronics mega-stores, even though these are often good places to kick the tires and do the actual purchase. Because so many of these stores exist, we can't address each one appropriately. Most of our thoughts on the national mega-stores apply to regional stores, too — we just don't have the space in this book to get into, for example, the chain of electronics superstores in Peoria, Illinois. (If there is one, we're sorry that we're not giving you details, Peoria!)

Note that HDTVs ship from their manufacturers optimally configured for the bright lights of retail stores. So often, when you're comparing HDTV images on TVs in a Best Buy or Sears, you're really looking at very extreme configurations that you won't be using at home. Be sure to check out reviews at some of the sites mentioned here, as well as your favorite hardware magazines. These reviews are done in a more apples-to-apples environment where the HDTVs are properly configured and more comparable with one another.

Be sure you know what you're buying. Many HDTVs don't come with actual HDTV receivers or tuners, meaning they're only HDTV-ready, and you have to buy a separate device to get HDTV programming. In some instances, a content provider such as a cable or satellite company provides the necessary tuner or receiver for free, so it helps to know where your content is coming from before you plunk down your hard-earned cash.

Crutchfield

Crutchfield (`www.crutchfield.com`) — an online retailer — stands head and shoulders above others for multiple reasons. It offers free shipping, free return shipping, and a 30-day trial of all equipment. Crutchfield also provides free "white glove" in-home delivery for larger HDTV models in which a professional delivery crew carries your TV into any accessible room in your home, places it on your stand or cabinet, and if you'd like, even takes the packaging away.

Crutchfield has a wide selection of HDTV options, organized for easy browsing. If you go to this site knowing what you want, you can get in and get out quickly.

If you go to this site still uncertain as to what you want to buy, plenty of information can help you make up your mind. The Crutchfield Advisor area includes hundreds of articles that provide you with answers to all sorts of HDTV questions. The site also includes glossaries, background information, and an option for live online chat to get a second opinion.

Above all, Crutchfield has a great reputation for knowing everything about what it sells and helping the customer through the whole life cycle of the purchase. This consultative approach has won legions of repeat customers. Crutchfield generally costs a little more (they sell most stuff for *MSRP*, or manufacturer's suggested retail price, rather than at a discount), but you get all this handholding in return. Crutchfield offers technical support for life, which, for newer technology such as HDTV, can be critical.

Buying an HDTV also means having to access HDTV sources, and buying everything from someone like Crutchfield ensures you don't have a weak middle HD link, like a video switch that doesn't switch true HD content. If you're going to buy more than an HDTV, think about at least talking to Crutchfield.com to see the big picture of everything you need to get for end-to-end HD in your home.

The pitfalls of online ordering

Ordering online is, in most ways, much easier than buying from a brick-and-mortar store. It is simply much easier to find the models you're seeking and to do oranges-to-oranges price comparisons. Even if you don't wind up buying a system online, you can use online price-comparison engines, such as the How Stuff Works site at `http://howstuffworks.shopping.com/xGS-hdtv~linkin_id-3055268` or the excellent PriceGrabber site at `www.pricegrabber.com`. What's more, in some instances, you can often avoid paying sales tax — which adds up very quickly on a multi-thousand-dollar item.

But you must consider these issues when attempting to order HDTV equipment online:

Shipping costs: These can wipe out savings, especially if the online vendor charges by weight instead of imposing a flat fee. Although many of the leading sites offer free shipping, at least for orders over a certain amount, some still charge a significant shipping fee. Most online vendors also expect you to pay return shipping, if needed, and many charge you a restocking fee if you return the unit. Most require that you return the system in its original packing and that you call and get return authorization before shipping it back. Vendors also vary in regards to the available forms of shipping, which can drive up cost.

Installation: Installing an HDTV system might not be rocket science, but it can be a headache to non-technical buyers — and a backache to the supergeek. *Remember:* Most online vendors simply ship the system to your house but don't carry the unit inside and set it up. Some retail stores (and a few online vendors) handle setup, but that adds to system cost. If a store is shipping the HDTV to you freight, you might be expected to accept the box right off the truck — try carrying that new 103-inch plasma up the driveway!

Warranty: Some manufacturers with in-home service warranties won't let you return the item after you've signed for it — it has to be repaired on premises. That means waiting for repair parts and scheduling time with repairmen. Yech. So check your box *very* carefully before you sign for the item.

Are they legit: This is, perhaps, the most important consideration. HDTV equipment makers want to make sure that only authorized online vendors sell their gear. The smart thing to do before making a purchase is to check the equipment maker's Web site to make sure your online vendor is authorized to sell that specific piece of equipment. You should see something like "Authorized Factory Dealer" displayed prominently on the site. Otherwise, you could wind up without a legitimate warranty on the system, which could cost you big bucks down the road. Unfortunately, not all manufacturers make this information readily available on their sites, so you might end up trading e-mails with customer support to find out whether your source is authorized.

Dell, Gateway, HP

Computer manufacturers are cashing in on the HDTV craze in a number of ways. Gateway Computers (`www.gateway.com`) first entered the home-entertainment market in some attention-getting ways. It was the first major retailer to offer a plasma-screen TV for less than $3,000. It also has led the market in its launch of home-entertainment gear built around computers. Dell, HP, and other computer manufacturers have followed suit with their own lines of plasma and LCD TVs. Indeed, the lines between a TV and a computer monitor are blurring rapidly — after all, a TV is just a monitor with some fancy innards for a tuner and some extra audio/video interfaces. As the tuner has moved out of the TV and into set-top boxes and PCs, the bulk of what is left is a monitor. Large-screen LCDs and plasma displays are increasingly common, as are home-priced, business-class, HDTV-capable projectors.

What to look for when shopping online

To even a trained eye, all online shopping storefronts can look alike. How can you make sure you don't get ripped off? Well, you can never be 100 percent certain, but here are some things to look for when trying to ascertain whether a site is "real." You want a site that_ meets these criteria:

- Accepts major credit cards
- Has visible and understandable policies for privacy, for returning items, for when credit cards are charged, and for getting in touch with the company
- Lists a physical address of some sort on the site
- Has a shopping-cart interface for taking and totaling orders
- Offers a secure server for the transaction (Look for a lock in the lower-right corner of your browser to show that a secure transaction is underway.)
- Has product-specific pages so you can make sure you're truly buying exactly what you ordered

Check out a seller's reputation. We recommend CNET.com as a good source for this sort of information. Whenever you click the Check Prices button to check the latest prices for a product from online merchants, you see a listing of pricing by vendor. The CNET Certified column tells you how well the company stacks up on CNET's ratings, and you can click the Store Profile link listed under each store logo in order to find out specifically what is driving each company's ratings. (We never buy from any store that has received less than 85-percent positive feedback from CNET's users.) For a benchmark example, check out Crutchfield's ratings at `http://reviews.cnet.com/4011-5_7-278703.html?tag=mlpmerch`. They rock!) Note that CNET's detailed Store Profiles are much more comprehensive than you'd find on other sites, like Amazon.com. Definitely make it a stop on your tour for detailed online store reputations.

Also seriously consider using a service like Escrow.com to hold your money until the product arrives in one piece. Escrow.com is easy to use:

1. The buyer and seller agree to the terms and details of the transaction.
2. The buyer sends payment to Escrow.com. Payment is verified and deposited into a trust account.
3. The seller ships the merchandise to the buyer, knowing that the buyer's payment is secured.
4. The buyer accepts the merchandise after having the opportunity to inspect it.
5. Escrow.com pays the seller after all conditions of the transaction are met.

An escrow service isn't free (what is?). For your peace of mind, you pay somewhere between 3.25 percent and 6.3 percent (depending upon the level of service you choose).

Finally, check out the restocking policy. These policies are generally different for a site based on whether the item was opened and whether it was defective. We've seen restocking fees from 0 to 30 percent, and the seller can charge shipping as well, depending on why you're returning the device.

For the computer manufacturers, HDTVs are a natural outgrowth of their media center PC efforts — where PCs act as your home entertainment device. Indeed, you can often find great bundling deals involving computers and big screen monitors. One great benefit of buying from Gateway, Dell, or another computer company is that it's easier to convince the accounting department at work (not to mention the IRS) that this is a valid business expense. We're not kidding. We know one guy (shhhh!) who had a $5,000 use-or-lose department budget for computer gear. He basically bought a $1,000 computer and a $4,000 "computer screen." You oughta see his presentations at home!

Computer manufacturers have competitive warranty and return policies with other retail options. The companies generally offer extended warranty options (which might be appealing if you're accident-prone). The more expensive option includes on-site repairs for accidental damage to the TV set — and even to the remote. The better computer and retail sellers offer on-site exchange services: If they cannot fix the problem over the phone, they ship a new TV to your home, install it, and take away the old one. They also have on-site installation services that include speaker and wall-mount installations. Pretty comprehensive for computer geeks.

Many online shopping sites (such as Amazon.com) also sell HDTV-ready sets bearing the familiar computer makers' logos, so don't be surprised if you run across those brands in those venues. However, we recommend that you always visit the manufacturer's home site so you can see what special deals and bundles are available there.

Sam's Club, Costco, Wal-Mart

If you're looking for a wide selection of all the best and leading-edge HD televisions, you wouldn't go to Sam's Club (`www.samsclub.com`), Costco (`www.costco.com`), Wal-Mart (`www.walmart.com`), or any of the other non-electronics outlets. You go to these places for one real reason — low prices. Their specialty isn't in offering a big selection, but rather it's in offering something of decent quality at a low price.

Danny got his first HDTV at Sam's Club. Because of the rate at which HDTV prices were plummeting, he didn't want to spend a lot of money on a great TV that would be half the price in another year or two. He figured he'd dabble in the technology, determine what he liked about various technologies, and make the real serious HDTV purchases when things had matured a lot more. (It's hard as writers who closely track technology to buy the current technology when we always know more cool and less-expensive devices are right around the corner.) You might want to do the same.

These outlets don't guarantee they stock or carry any specific television. If you're looking for a specific model that got a great review in *Home Theater Magazine,* you might or might not be able to get it in these stores. It's hit or miss.

What really separates these stores from each other are the options for delivery and installation. These options range from "cash and carry" to full in-home delivery, installation, and removal of the trash. Wal-Mart, for instance, helps you carry it to your car or delivers the product curbside to your home. The downside to Wal-Mart is it doesn't have any programs, at least at the time of this writing, for in-home installation. Costco has a deal with Installs, Inc. (`www.installs.com`) to come to your home and take care of the whole installation process for you. (You still have to get the TV to your home and in your living room, though). An HDTV delivered by Sam's Club "will be unloaded and carried to your first available dry area or exterior location of choice but no further into your home than across the front threshold." These delivery options change over time, so just make sure you check out your options if you own a Mini (or simply don't want to risk dropping your new HDTV in the driveway).

Many online and retail outlets offer extended warranties on your HDTV purchase. These are hard to justify sometimes because they can cost up to about one third of the purchase price. In a steady-state price market, this might make sense; in a rapidly decreasing and technology-advancing market, as expensive as it seems, it might not make sense. When the first edition of this book came out, a 40-inch plasma-screen HDTV cost $3,500 with an extended warranty cost of about $1,200. Today, $1,200 can buy a whole new plasma HDTV.

Still, if you aren't one to roll the dice on whether your TV will break (or you have a clumsy friend you're convinced will spill some latte on your electronics), here are some things to look for in extended warranties:

- If they claim 24-hour service, they typically are talking about having someone to talk to 24 hours a day, but their repair centers usually operate under normal business hours.
- Check the fine print about *where* the service takes place. Most reference the original warranty, and if the original warranty is not on-site, then the extended warranty probably is not on-site either. If you are looking for in-home service, make sure that's what you are buying and extending.
- Many consider their obligation fulfilled if they provide you with a replacement product with similar features, because the model you purchased might not be available any more. So realize that your extended warranty doesn't guarantee that your specific model will be retained if the screen goes bad or some other catastrophic issue comes up.

- These plans have all sorts of variations due to the state in which you reside. For instance, under Wal-Mart's extended warranty service, if you live in Arizona, you can get a pro-rated refund if you cancel the extended policy for some reason; if you're in Connecticut, you can get your warranty period extended for the time that your TV is at their shop being repaired. Check state-level exceptions carefully.

Above all, know that not all extended warranties cost the same. You can find vendors who offer low-priced HDTV sets but higher-priced extended warranties in order to make up the difference. If you plan to get a warranty, shop around for the total HDTV+Warranty price.

What about Amazon.com, eBay.com, and CNET.com?

When you think about buying anything online, Amazon.com and eBay.com certainly come to mind. For people with a more technical slant, CNET.com comes to mind, as well. These sites have expanded their horizons immensely from their book, auction, and technical roots, respectively, to encompass a huge legion of sellers of all sorts of goods — including HDTVs. With the exception of some models shown on Amazon, these sites mostly serve as price-comparison engines, showing you where you can buy the units displayed for the least cash — that is, they aren't direct retailers themselves.

Amazon.com has a special section for buying HDTVs, which you can find by typing HDTV in the search box on the main Amazon.com page. There you can see direct deals that Amazon has for selling the items under its own storefront, as well as links to major vendors like Tiger Direct and J&R Music and Computer World. As with most of Amazon's entries, you get detailed information about the product, customer reviews, and (here's what we like) what other items people looked at when they were looking at this product. Most HDTVs have a bundled purchase option with some sort of mounting product, too.

CNET (`www.cnet.com`) has really emerged as a powerful force in comparison shopping as well, with content deals with all the major players to show pricing and inventory on the CNET site. CNET doesn't sell anything itself, making it a truly independent source of information. What's powerful about CNET is that it offers its own reviews of each product, and reflects users' opinions, as well. CNET.com should be a stop on your HDTV hunt.

As far as eBay is concerned, this is a more unpredictable adventure. We priced one 60-inch plasma on eBay and found that pricing for the system — brand new, in the box — ranged from $2,200 to $17,999. It's hard to explain such a price range, but definitely think twice before you send that much money to someone on eBay (or any stranger). Be aware that many of these vendors might not be authorized vendors, which could mean a voided warranty. *Caveat emptor* (buyer beware!), especially with newer vendors who don't have a long track record. Use CNET.com to cross-check vendor ratings. Still, you can find valid, great HDTV deals on eBay, as with many of the other items sold on the site.

Circuit City

Circuit City (www.circuitcity.com) combines online ordering with in-store sales and does a pretty good job on both fronts. Circuit City online boasts a wide range of products, and you can see what products are available in your area. These products can be compared (up to five items) in a side-by-side format on the Circuit City site.

One major bonus is that the company offers free delivery and setup, and for an extra fee, even connects other video devices within your home to the HDTV system. The vendor also offers immediate advice on the Web site about which accessories are required for which systems, making that part of the purchase easier, as well. Customer ratings are provided along with the listings of the HDTV units, but they don't seem to vary too much from set to set.

More detailed discussion of HDTV is found in the Home Theater Headquarters, where Circuit City has a section on Understanding HDTV and the HD Experience. This area isn't very detailed though (especially considering that you were smart enough to buy *HDTV For Dummies!*). The best articles are actually from CNET.com and *Home Theater Magazine*.

Circuit City also offers in-store sales of HDTV units, which presents a perfect opportunity to see in advance what you're buying if you're close to a Circuit City store. You can even order online and pick up the system at the nearest Circuit City store and return it there if need be.

Best Buy

Like Circuit City, Best Buy (www.bestbuy.com) combines features of a brick-and-mortar store and an online service, capable of price and feature comparisons. Best Buy's online interface offers the option of comparing related systems by features and price. Whether you can actually buy the system online is tied to your zip code — not all systems are available even in sites geographically close to Best Buy store locations.

Best Buy has an extensive selection of HDTV systems. The company offers free shipping using its Retail Delivery service, which means your purchase is shipped from its stores, but the service is more narrowly defined than Circuit City's similar option. You must confirm an address that falls within the free shipping area, and the free shipping doesn't cover all items. If a specific free shipping option isn't designated, the only way to know for sure whether you qualify is to put the item in your shopping cart, enter the appropriate information, and see what Best Buy says. It's possible to pick up the item at a Best Buy store and return it there. Best Buy lists the required accessories and often offers a discount on these items with the HDTV system.

Best Buy's home delivery service includes delivery into the room of your choice. They unpack the item and, if requested, dispose of the packaging. For an additional cost, they can provide installation of other home entertainment components.

ClubMac

ClubMac (`www.clubmac.com`) originally opened to sell — guess what — Apple Mac products. It has expanded its mission to a wide variety of other electronics. So don't be fooled by the Mac-dominated home page. The store meets all the baseline characteristics for online selling. It also offers the widest possible selection of delivery and payment options (including C.O.D.), as well as a toll-free number for customer service and online order tracking. The company's HDTV selection includes a reasonable variety of manufacturers and pricing options, as well as an extensive set of cables and accessories.

Microsoft, Apple

With the convergence of the PC and entertainment, Microsoft isn't going to be left out of the action. Microsoft's online catalog (`www.microsoft.com/windows/catalog/default.aspx`; click the Hardware tab) includes information about vendors that enable a personal computer to become an HDTV receiver. These receivers connect to the PC via USB ports or through internal cards that convert a PC to a receiver. This is cutting-edge stuff and not for the technically skittish.

Apple (`www.apple.com`) is increasingly supportive of HDTV, too, as it has moved from computers into your living room. Apple's LCD screens are nothing but stunning — the 30-inch Cinema display has a whopping 2,560-x-1,600 pixel resolution. Wow! Apple's iTV player is just coming out as we write this book, but from what we've seen, it's a product we want in our living rooms. Apple wants to make buying, playing, porting, and storing HDTV easy with iTunes, iPods, iTV players, and more. Even if you're a Windows bigot, you need to see what Apple has in this space because it's bound to be sexy, integrated, and fun.

Sony Electronics

This option is a limited one. The company has large retail Sony Style stores in New York and San Francisco, and smaller ones in Chicago, Southern California, Boston, and a couple dozen other places in the United States. But Sony is specifically focusing on high-end items such as HDTV units for its retail push. Sony is probably not the cheapest option, but if you're not too price-sensitive (or you're enamored of the Sony brand) and your address has the right zip code, it's an option.

Don't leave home without it

Expensive HDTV electronics are a good reason to get a Platinum American Express Card if you don't have one. America Express offers

- **Return Protection:** If your retailer won't take an item back on return within 90 days of purchase, Amex reimburses you up to $300 per item ($1,000 total per account per year). This might not cover your HDTV, but it could cover that up-converting DVD player that didn't quite work for you.
- **Purchase Protection Plan:** If your HDTV is stolen, lost, or accidentally damaged (and that includes vandalism), you're covered. The coverage is limited to $10,000 per occurrence, up to $50,000 per cardmember account per policy year, and is "in excess of other sources of indemnity," which means your own insurance gets the first crack at this reimbursement.
- **Buyer's Assurance Plan:** Amex also extends your warranty by up to a year (double the original warranty if it's a year or less, or one year for warranties that are more than a year but less than 5). The limits here are that "The Buyer's Assurance Plan won't pay more than the actual amount charged to your card for the item or $10,000; whichever is less, not to exceed $50,000 per cardmember account per policy year."

Given the extra benefits the American Express card conveys, this card is worth getting your hands on.

Chapter 26

More Than Ten HDTV Accessories

In This Chapter

- Shaking it up (baby!) with transducers
- Saving your marriage by going wireless
- Showing your HDTV movies in the backyard
- Going Hollywood all the way

No HDTV is an island! It should be connected to and surrounded by accessories that make the HDTV but one part (a big part) of the overall viewing experience. From wireless headphones to vibrating chairs, you can extend your HDTV experience in all sorts of ways.

Kicking Some Butt (With Transducers)

One of the best scenes in the movies ever, we think, was in *Jurassic Park,* when Dr. Alan Grant (Sam Neill's character) and his entourage encounter the Tyrannosaurus Rex for the first time, in the unsecured open terrain of the park. Who can forget the cup of water vibrating to the footfall of the beast? Scary!

In the movie theater, the audience felt that through the incredible bass management of the speaker systems. And in a home audio system, the special effects audio channel on the DVD drives some of that through the subwoofer, too.

But none of that compares to what you can get with audio transducers from companies such as Clark Synthesis (`www.clarksynthesis.com`) and The Guitammer Company, Inc. (`www.thebuttkicker.com`). With these transducers, you can get one giant-T-Rex-step closer to the ultimate HDTV theater!

Transducers are marketed under many names — *bass shakers, tactile transducers,* or as one vendor calls them, *buttkickers.* The units are screwed onto the bottom of your furniture or into the frames of your floorboards. They're better than subwoofer-based effects because they can be very localized — only your couch shakes, for instance, if they're attached to the couch's underframing. (Your neighbors will be happy about that!)

If you live in a house that has a lot of concrete, stone, and brickwork, you definitely want to look at adding a transducer to your furniture because these home-building materials don't conduct the bass very well. Even if you have a great subwoofer, you won't get as much low-frequency effect because of the room's materials. Spend the extra dollars on a transducer in these instances.

As with most parts of your HDTV system, you can spend a lot or a little on transducers. One aspect depends on how many you install. You can install one, two, or three transducers to a couch and get an increasingly better effect. With three across the bottom of your sofa, you can power the signal for the middle transducer from the LFE/subwoofer out of the processor-receiver and drive the two side ones from the left and right front channels, respectively. This gives you feedback across the range of the front speakers. If you're watching *Black Hawk Down,* when there's an explosion on the left, you hear the sound from the left channel as well as feel the lower frequencies from the transducer. If a tank rumbles across the screen, it rumbles across the couch, as well.

The lower-budget transducers have a fairly decent price difference. The Aura Bass Shakers (`www.aurasound.com`), which do a fairly good job, cost around $60. The higher-end versions that have more power and precision, such as the Clark Synthesis or Guitammer Buttkicker models, run from $200 to $500.

The units differ as to how high a frequency they support (that is, for how high a frequency they react to by vibrating). Some products support frequencies up to 800 Hz. We advise that you keep your frequency range on the lower end — between 5 Hz and 200 Hz. Otherwise, it seems the transducer is always rumbling along in the background, and that gets annoying after a while. You can control the frequency sent to your transducer via an equalizer, such as the $120 Audio Source EQ.

If you mount transducers directly to your furniture, you can get rubberized, molded mounts for your chair or couch legs that isolate the noise and vibration to your furniture and from the whole house. For instance, check out Clark's IsolationFeet at `www.clarksynthesis.com/home-feet.php`.

Read the manufacturer's recommendations closely for info on how to power each unit. Here are our recommendations:

- Use one amplifier per transducer, or at least no more than two.
- Lower-end models might not need anything more than a 20-watt amplifier; higher-end versions might need an amplifier of at least 100 watts.
- Often, transducers come bundled with an amplifier (active transducers). Consider getting one of these if you have questions. Aura Systems, for instance, has a Bass Shaker Plus bass enhancement package (AST-3B-4, $224, `www.madisound.com`) that includes a 100-watt digital amp, remote level control, bass control, and two pro shakers.
- Look for amps that have their own volume control so that you can tune the effect relative to the audio level. Don't worry about whether it's a high-quality amplifier — its use in this application isn't high fidelity.

Motion Simulators

Shaking things up with Buttkickers and the like is fun, but the rides at Disney World have progressed quite beyond the mere shaking your groove thing. You move up, down, left, right, forward, and backward — all the while it shakes you up.

For the vast majority of people — that is, those who don't have millions to splurge on their own amusement parks with Disney World–like rides — such an intense experience has been out of reach. Until the D-Box.

Unlike bass shakers, which provide only vibration or "shaking" in response to the audio track and are in reality merely transducers that vibrate rather than move air, the Odyssee motion simulator from D-Box (`www.d-box.com`) is a sophisticated motion-simulation system that lifts seating and occupants on three axes (X, Y, and Z axes, for pitch, roll, and yaw) at up to 2Gs (think F-14 at full throttle) of acceleration.

Odyssee provides dramatic motion that is precisely synchronized with on-screen action, which draws in viewers even more by allowing them to accurately experience the accelerations, turns, and jumps that they could previously only imagine. When a car rounds a corner in a 007 chase scene, turn with it. If the *Top Gun* jet fighter suddenly moves into a climb, climb with it.

Here's how it works: The basic system includes an Odyssee Controller and a set of Odyssee Actuators. The controller manages the translation of motion cues from the DVD's content to the motion of the actuators. The controller's microprocessors direct the vertical and horizontal movement of the actuators. Not all movies are supported by the system; codes must be programmed for each specific movie. Hundreds of movies and TV shows (like *Lost* and *24*) are supported now (see the list at `www.d-box.com/en/codes/index.html`), with more coming each month.

Just about everyone who's tried this system loves it. Until recently, this system was only for the rich and famous — the price tag for an entry-level system runs $15,000 — but a new certification program enables manufacturers of home-theater seats to put the two-axis actuators into the seating. Initially at around $5,000 a chair, this program promises to dramatically reduce the price over time. Check in with D-Box to see how cost effective it's gotten!

Power Conditioner

Most people don't know it, but the electrical power in your home fluctuates all over the place. Every time your refrigerator or air conditioner turns on, there's a surge in the power line to compensate. As devices turn on and off in the house (and in the neighborhood), power levels likewise ebb and flow with the current.

The bottom line is that electrical fluctuation affects your HDTV gear. You absolutely must protect your HDTV system with some form of surge protection. But surge protection protects you only from major changes in your electrical lines — you still could have issues with the consistency of the current level going to your expensive gear. For example, if your voltage drops, it might actually lower your amplifier's output. Ground issues can cause hums in your audio and lines on your video display.

Consider buying a power conditioner, which improves and stabilizes the AC power for your HDTV system. Power conditioners use various techniques to restore your AC power to a true 60 Hz, 120-volt signal, which can offer better audio and video performance.

Many of the major stereo cable and wiring vendors offer power conditioners. One of the better-known vendors is Monster Cable, which sells some really cool (and cool-looking) Home Theater Power Centers (`www.monstercable.`

com/power). These products provide surge protection, voltage stabilization, and noise filtering. The more expensive ones even have a neat digital voltmeter readout on the front. (We insist on cool readouts on all our gear!) Expect to typically spend from $100 to $500, to potentially upwards of $1,000, for a power conditioner. Protect your system — get one.

DVD Changer Controllers

If you're anywhere near the movie aficionados that we are, you have hundreds of videotapes and discs littering your home. What used to be a contest for bragging rights with the neighbor over who bought the most discs at Wal-Mart has grown into a mess.

The DVD changers we discuss in Chapter 11 are great for smaller collections, but what do you do with the larger collections that you want to be able to access centrally from anywhere in your home?

From the same company that makes the best audio server in the industry comes a video option based on the same technologies. ReQuest Multimedia's VideoReQuest (www.request.com) enables you to control up to four Sony DVP-CX777ES 400 disc DVD changers through a simple on-HDTV-screen interface. Imagine being able to instantly select from any of 1,600 DVDs in a collection with a simple, intuitive, on-screen user interface.

You can access DVD movies by title, genre, MPAA rating, actor, and director. VideoReQuest provides for automatic discovery of DVD information by using Internet lookup, too — so no more pecking at a keyboard entering titles (what a pain!). Just load the DVD into the system, and VideoReQuest takes it from there.

What's more, you can integrate your VideoReQuest into your home control network so you can start movies from touchscreens located throughout the home. If you have a home integration system (like Crestron or AMX), VideoReQuest can connect via RS-232 or Ethernet interfaces. The system outputs to any VGA, component-video, S-video, or composite-video port.

Simple. Functional. Clean. A perfect solution to a growing problem. The VideoReQuest retails for about $2,500. But we also highly recommend having an AudioReQuest in your audio system as well to store all your CDs.

Show HDTV Outdoors

There's nothing like an outdoor HDTV movie. Starry skies, fresh popcorn, a 10-x-20-foot screen, and lots of friends and blankets. It's easy to do, using common items found at Home Depot and your nearest party tent vendor.

In our experience, the issues with HDTV outdoors are less about having a pristine audio and video experience, as they are about being able to show a huge screen with audio everyone can hear — and picking a good movie. The sheer fun far outweighs any issues that are much more noticeable in an enclosed room built specifically to be a theater.

Your outdoor experience requires these basics:

- **A video projector that you can take out to the backyard:** Because of the greater amount of ambient light, look for projectors with at least 1,200 lumens, preferably as much as 2,000 lumens. You can now find these projectors for less than $1,000.
- **A video source:** You can use either a portable DVD player or a wireless link back to your HDTV system.
- **Speakers:** You can just run long speaker cables from your source or pick up some wireless speakers to go the distance. We've used some off-the-shelf 5.8 GHz speakers from Radio Shack that work just fine — it changes models occasionally but usually it has a pair that costs around $200. Soon, you'll be able to get 802.11-standard gear in most major computer stores at similar prices.
- **Screen:** You can make an outdoor video screen with the same stuff that party companies use. You can make a screen for $300 to $500 (everything but the pipe on the following list totals about $150):
    - About 30 10-foot pieces of 1-inch EMT pipe from a home store

      This pipe usually totals $150 to $350.
    - A 10-x-20-foot party-tent tarp
    - Party-tent junction points for the corners
    - A bunch of ball bungee cords to mount the tarp

If you're showing video outdoors someplace damp, cover your gear with a blanket before the show — the evening dew can damage your electronics.

If you have a lot of outdoor shows, consider buying a projector that's optimized to go outdoors. You can get pretty good projectors for less than $1,200. When it isn't in the backyard, you can project it against a screen in your living room or your office. You can check out the latest models that are ideal for the outdoor experience — and get ideas for how to do this — at our outdoor home-theater site at www.digitaldummies.com/projects/outdoor.asp.

You can also get outdoor screens that are inflatable, ranging from a low of $200 for a 12-foot screen at Wal-Mart (www.walmart.com) and reaching into the thousands of dollars for commercial-grade inflatables such as the Airscreen (www.airscreen.com, starts at $8,000). We like the hard frame options better for more stability and a more taut screen. It's more resilient in the wind, too.

Creature Double Feature in 3D

Most people have rather private experiences with 3D, dating back to drive-in movies when they were kids, or relating to some of the stunning latter-day movies from IMAX. Some people have been on the *T2: 3D* "ride" at the Universal Studios theme park, or the *Honey, I Shrunk the Kids* exhibit at EPCOT in Disney World.

Regardless of your experience, it's hardly been a common part of your entertainment regimen.

For years, the most common 3D technique used has been the anaglyph, a means of creating a single image from two color-coded images that are superimposed on one another, giving you the sensation of depth perception. The red/blue cheap cardboard glasses are the anaglyph experience that many people equate to 3D.

However, a smart Canadian firm called Sensio (www.sensio.tv) has taken a different path and created the industry's only high-quality 3D system. Sensio uses a method wherein glasses with electronic LCD shutters alternate left and right, allowing each eye to view the screen every other sixtieth of a second. Each eye sees only its corresponding image (the left eye sees only the left angle images) on-screen. The Sensio S3D-100 system is a base unit that is another component in your system. The base unit is connected between your DVD player and your display device. You put your special Sensio-encoded DVD in your DVD player. The S3D-100 reads the signal coming from the DVD player and transmits an alternative (left and right) progressive image to the screen and also sends a signal over an IR emitter to the wireless IR-driven glasses. The IR has a range of 20 to 30 feet in a room.

You can't use the Sensio box with just any TV, DVD player, or DVD. You need a special DVD that carries the Sensio encoding. (There are almost two dozen such DVDs now.) You also need something that can take in a VGA connector — the S3D-100's output is analog RGB via a VGA connector, which is common on most video projectors. Most HDTV sets have component video on board — you can handle these with a VGA-to-component converter such as one from Key Digital (the $375 KD-VA5).

You also need a DVD player with component-video outputs. The Sensio processor requires an NTSC signal, which means you must set a progressive-scan DVD player's output to *interlace* mode. (If you need to know more about progressive and interlace mode, check out Chapter 21.)

You can't use the Sensio with a plasma TV because the plasma screens emit a great deal of infrared, which interferes with the glasses' interpretation of the sync signal from the emitter.

Currently, Sensio is trying to expand its cadre of 3D movies. The 250+ 3D movies that have been made since about 1915 were displayed with the anaglyph process. If you want to watch those, you need a special set of anaglyph glasses; they're useless on a Sensio system. Sensio is presently working with the major Hollywood Studio to convert those movies in the Sensio 3D format. The popularity of some more recent movies — *Ghost of The Abyss (Titanic), Shrek 4D, Spy Kids 3D,* and other 3D movies — gives 3D fans like us hope for the future. At $3,000 per box, we think the Sensio system is a must-have HDTV accompaniment.

Because the glasses shutter on and off rapidly, Sensio warns that folks who are prone to seizures should avoid wearing the glasses.

Wireless Headphones

There's nothing like a great action-packed movie such as *The Italian Job* screaming over the HDTV. That is, until your spouse leans over and asks you to *turn it down* in no uncertain terms! Wireless headphones not only help you hear your HDTV as loud as you want it, but they also save marriages, too!

Wireless headphones also give you mobility — you can get up and let the dog out, and not miss any action.

These aren't hard to install or use. Wireless headsets use radio to connect to a base station that's connected to your audio receiver.

There's a huge variance in headphones on the market. Some fit over your whole ear and others fit in your ear. Some headphones deal with ambient noise and use DSPs to process Dolby Digital 5.1.

Dolby has done wonders with this technology application. No matter how many channels or speakers you have in your HDTV setup, you still have just two ears through which all the audio is processed. Dolby Headphone (`www.dolby.com/dolbyheadphone`) tries to reproduce what is arriving at your ears by using digital signal processors to simulate Dolby Digital 5.1 over the two headphone channels. Dolby Headphone offers three listening modes, each based on acoustic measurements of real rooms. DH1, or Studio, is a small, well-damped room appropriate for both movies and music-only recordings. DH2 (Cinema) is a more acoustically "live" room, especially suited for music but also great for movies. DH3 (Hall) is a larger room, more like a concert hall or movie theater. All products equipped with Dolby Headphone include DH2. On the products offering two or all three environments, you can easily switch between modes to suit the material and your own preferences.

If you plan on using headphones a lot, see whether your receiver supports Dolby Headphone encoding. We highly recommend it.

What headphones do we use? A lot of the wireless headphones geared for the television are outright lousy. The only decent wireless headphones we've run into are made by Sennheiser (`www.sennheiserusa.com`), specifically the RS65 ($190) and the RS85 ($250). The RS85 sports a high-performance noise-reduction circuit that gives it better performance at the edge of its operating range. Though pricey, both models work great, and their signals really travel far in a home. What's more, you can get an extra headphone for each of the models, so Mom and Dad can both watch loud movies while the kids sleep. (Kill that, Bill!)

Check out HeadRoom at `www.headphone.com`. It's a great online resource/store for the latest and greatest in headphones, including some good reviews and information. That's where we shop for 'phones. (No one paid us to say that either!)

In Search of That Great Remote Control

There are soooo many remote controls on the market, it's daunting just to talk about a few. Remote controls have evolved a lot since the earliest universal remote controls hit the market, allowing you to control multiple devices from one remote control.

A quick scan of HDTV Web sites today shows all sorts of remote control options: tiny, large, color, touch-sensitive, voice-controlled, time-controlled, and on and on. You can spend $12 on a great remote, or $4,000. You can keep it confined to your HDTV theater or go whole-home.

Types of remotes

Typically, remotes fall into the following categories, presented here in increasing order of functionality and desirability:

- **Standard/dedicated remotes:** These are the device-specific remotes that come with each device.
- **Brand-based remotes that come with a component:** Brand-based remotes work with all sorts of devices from the same manufacturer.
- **Third-party universal remote controls:** Universal remote controls are designed to work with any electronics device by way of on-board code databases.
- **Learning remotes:** Learning remotes can learn codes from your existing remote controls. You simply point your remotes at each other, go through a listing of commands, and the remote codes are transferred from one to the other.
- **Programmable remotes:** Programmable remotes allow you to create *macros,* sequential code combinations that do a lot of things at once.
- **Computer-based remotes:** Computer-based remotes take advantage of the growing number of computing devices around the home to provide remote control capabilities. This category includes PDAs, Web tablets, portable touchscreens, and PCs.
- **Proprietary systems:** A number of closed-system control devices enable you to integrate control of all your home-theater devices on a single control system. Companies such as Niles Audio (`www.nilesaudio.com`) and Crestron (`www.crestron.com`) are renowned for their control systems.

Remote controls can use infrared (IR) or radio frequency (RF) to communicate with their base device(s). Some remote control systems use either tabletop or wall-mounted touchscreen displays.

Two new classes of remotes include two-way operation and voice control. With two-way operation, higher-end remotes can interact with the controlled unit to determine its *state,* so you can determine whether a unit is on or whether the input is already set to Video1 before changing that state. And with two-way operation, you can check your actions to make sure they were carried out, too.

Voice control enables you to talk to your remote control to effect changes in your HDTV environment. Want the lights dimmed and curtains opened? Just say, "Start the show," and a preset sequence of activities begins.

More and more remotes have a *docking cradle* where you can charge your remote, access the Internet for revised programming schedules, or update its internal code databases.

Sexy remotes

The more you spend on your entertainment system, the more you'll probably spend on your remote controls.

These are some of the best remotes on the market:

- **Pronto (Philips; $199 to $1,200; `www.pronto.philips.com`):** Philips has a solid line of remote controls that have defined the leading edge of home-theater remotes in a lot of ways.

  The $1,200 fully programmable and learning Pronto Professional TSU9600 wireless home control panel combines the best of two infotainment worlds: home-theater system control and 802.11 wireless broadband Internet access. The LCD display offers an exceptional full VGA 640-x-480 resolution on a standard 3.7-inch diagonal touchscreen. The touchscreen is complemented with 24 hard buttons (backlit by blue LEDs), including a five-way menu cursor, five custom labeled buttons below the display, and a rotating ring for use with media servers. With an optional IR/RF capability, the unit allows you to control infrared devices in cabinets or in other rooms with the simple addition of a Philips receiver.

  You can also use the built-in wireless system with Escient music servers and Lutron lighting systems. The remote is equipped with a dedicated user interface to control Lutron RadioRA lights and shades and, in turn, provide feedback on the remote's 3.7-inch VGA screen on the activity performed. Mood settings enable the complete room setting to be illuminated according to the desired mood at the touch of a single button — you don't need to control individual dimmers and switches.

- **Harmony (Logitech; $199 to $499; `www.logitech.com/harmony`):** Harmony has been leading the pack toward intelligent, learning, and programmable remotes. The company has numerous products, so you might have a hard time figuring out which is best for you. The $250 Harmony 895 Advanced Universal Remote Control uses both radio frequency and infrared wireless signals to control multi-zone home-entertainment systems and the most advanced home-theater setups. These remotes are focused on actions, not devices, such as listening to a baseball game or watching a video. Harmony's remote has a color LCD screen, backlit buttons, and links to your PC via a USB connection to program the remote to tie together multiple actions at once, in order. So you can turn on the TV, the receiver, and the amplifier, and set them all to the correct settings, with one click. Very nice. Harmony was bought by Logitech in 2004, and we LOVE Logitech's gear.

  At the time of this writing, Logitech's latest remote control and the company's first touchscreen product is the $499 Harmony 1000. Wow! With a 3.5-inch color TFT LCD screen, control of up to 15 devices, an optional RF extender, and the company's special activity-based online programming, this remote sets a new standard for controlling your home entertainment at this price point. The 1000 has all the activity-orientation of the other Harmony products, but with a more high-end touchscreen look and feel. At a much lower price point than the Philips TSU9600, we think the 1000 will fit into more homeowners' budgets.

- **weemote (Fobis; $24.95; `www.weemote.com`):** This is a remote designed just for kids ages 3 to 8, so you can limit the channels they watch and make them responsible for their own remotes (in other words, keep their paws off yours). The company is into its seventh generation of product with a completely redesigned model. It can now learn selected button controls from your existing remotes, provide a one-step setting for DIRECTV users, streamline programming steps, and other cool things. The best part is that it's optimized for small hands and kid programming control.

  Fobis has also developed a weemote, sr., that's not as cutesy as the kid's remotes, but is designed to be incredibly easy to use for seniors or folks with vision problems or decreased hand coordination.

- **Your cell phone as a remote control:** Most Palm and Microsoft Mobile OS-driven smart phones and similar PDAs have an IR port, and you can download any of a number of programs to customize your home-theater environment. These usually cost from $20–$35, and can be found at places like PalmSource (`www.palmsource.com`) or ZDNet's Mobile Downloads (`downloads-zdnet.com.com`).
- **Touchscreens galore around the house:** Crestron (`www.crestron.com`) rules the upper end of touchscreen options. Crestron's color touchpad systems are to die for, or at least second mortgage for. Other options include HAI's OmniTouch (`www.homeauto.com`) and up-and-comer Control4 (`www.control4.com`). The touchscreens can be all over your home and allow you to control your HDTV from anywhere. If your HDTV is fed into a whole-house audio system, you'll certainly want to change the channels every once in a while from another room so you can start listening to *The Closer* from the first frame, if you can't be in front of the TV to watch it — if you miss the first five minutes of an episode, you're lost for that episode!

Remote control technology advanced quickly. For remote control options, visit Remote Central (`www.remotecentral.com`). It has great reviews and tracks the newest remotes.

Keep your eyes on this company: Hillcrest Labs (`www.hillcrestlabs.com`). This small startup is shaking things up by proposing to get rid of the remote control in favor of a six-dimensional airmouse with just two buttons and a scroll. It senses your hand's movement and translates that into a cursor on the screen that can click on-screen buttons — instead of physical buttons on a remote control. We've used it and love it. We'd happily give up our best remote control for a Hillcrest-powered solution. You'll start seeing Hillcrest's technology appear more and more as a part of your TV life as manufacturers roll out Hillcrest-based TV air controls and on-screen interfaces. Finally, something we can use in the dark or without our glasses on!

Bring Hollywood Home

If you really want to show off your HDTV, you can build out your HDTV setup with all the accoutrements of a true theater:

- **Popcorn machines:** Complete with the swing-down popping bucket, these $400-to-$900 machines give you freshly popped popcorn and that movie-theater smell. Check out www.popcornpopper.com for some cool models.

  Go for a 6- or 8-ounce kettle if you can; the 4-ounce kettle just doesn't make enough popcorn for a good-sized crowd, and you find yourself making more.

- **Candy stands:** You can get fully lit concession stands for your home theater, just like the ones found in real theaters. You can also get concession signs and candy bins. Check out places such as www.candyfavorites.com.
- **Personalized DVD intro:** Open up each showing with your own customized one-minute video, just like those in the theaters. (Your DVD player must be able to play DVD-R.) For dropping $150 or so at www.htmarket.com, you can personalize one of six home-theater intros, from an awards-night theme to a classic popcorn theme. How fun is that?
- **Furniture:** From full-sized ticket booths to Roaring '20s–style bar furniture, you can re-create almost any environment in your HDTV theater. The sky's the limit on pricing for these items. A ticket booth costs $2,000.
- **Theater-style roping:** You can get that velour-rope-and-stainless-steel-post look if you want — expect to spend around $300 a post and $100 a rope for the really nice models. Think about being able to make your mother-in-law wait in line to see the movie at your new home theater! We love that!
- **Themed carpet:** You can buy carpeting that's festooned with images of cinema stars, film reels, popcorn, and other themes. You can expect to pay up to $50 per square yard or more (as with any carpeting, this can vary greatly).
- **Themed theater lighting:** Cool wall sconces, tabletop lights, and standing lights add the theater look. Budget about $200 for each of these.
- **Film posters:** Complement your HDTV room with the latest movie posters. These are usually a standard 27-x-40-inch size, and cost about $10 to $20 for current films. Go online at www.allposters.com for current films, or get vintage posters at places such as www.moviemarket.com.

You can find these items and more at places such as www.hometheaterdecor.com and www.htmarket.com. Also, check your local Sam's Club or Costco — more and more of these stores are carrying vending machines, popcorn poppers, and other such accessories for your home theater. Finally, don't forget to check eBay for fun stuff — the Collectibles and Memorabilia areas often have old authentic theater items at good prices.

Index

• Symbols and Numerics •

• A •

• B •

• C •

• D •

• E •

• N •

• O •

• R •

• S •

• T •

• U •

• V •

• Z •

Notes

Notes

SINESS, CAREERS & PERSONAL FINANCE

0-7645-9847-3

0-7645-2431-3

Also available:

- Business Plans Kit For Dummies 0-7645-9794-9
- Economics For Dummies 0-7645-5726-2
- Grant Writing For Dummies 0-7645-8416-2
- Home Buying For Dummies 0-7645-5331-3
- Managing For Dummies 0-7645-1771-6
- Marketing For Dummies 0-7645-5600-2
- Personal Finance For Dummies 0-7645-2590-5*
- Resumes For Dummies 0-7645-5471-9
- Selling For Dummies 0-7645-5363-1
- Six Sigma For Dummies 0-7645-6798-5
- Small Business Kit For Dummies 0-7645-5984-2
- Starting an eBay Business For Dummies 0-7645-6924-4
- Your Dream Career For Dummies 0-7645-9795-7

ME & BUSINESS COMPUTER BASICS

0-470-05432-8

0-471-75421-8

Also available:

- Cleaning Windows Vista For Dummies 0-471-78293-9
- Excel 2007 For Dummies 0-470-03737-7
- Mac OS X Tiger For Dummies 0-7645-7675-5
- MacBook For Dummies 0-470-04859-X
- Macs For Dummies 0-470-04849-2
- Office 2007 For Dummies 0-470-00923-3
- Outlook 2007 For Dummies 0-470-03830-6
- PCs For Dummies 0-7645-8958-X
- Salesforce.com For Dummies 0-470-04893-X
- Upgrading & Fixing Laptops For Dummies 0-7645-8959-8
- Word 2007 For Dummies 0-470-03658-3
- Quicken 2007 For Dummies 0-470-04600-7

OD, HOME, GARDEN, HOBBIES, MUSIC & PETS

0-7645-8404-9

0-7645-9904-6

Also available:

- Candy Making For Dummies 0-7645-9734-5
- Card Games For Dummies 0-7645-9910-0
- Crocheting For Dummies 0-7645-4151-X
- Dog Training For Dummies 0-7645-8418-9
- Healthy Carb Cookbook For Dummies 0-7645-8476-6
- Home Maintenance For Dummies 0-7645-5215-5
- Horses For Dummies 0-7645-9797-3
- Jewelry Making & Beading For Dummies 0-7645-2571-9
- Orchids For Dummies 0-7645-6759-4
- Puppies For Dummies 0-7645-5255-4
- Rock Guitar For Dummies 0-7645-5356-9
- Sewing For Dummies 0-7645-6847-7
- Singing For Dummies 0-7645-2475-5

TERNET & DIGITAL MEDIA

0-470-04529-9

0-470-04894-8

Also available:

- Blogging For Dummies 0-471-77084-1
- Digital Photography For Dummies 0-7645-9802-3
- Digital Photography All-in-One Desk Reference For Dummies 0-470-03743-1
- Digital SLR Cameras and Photography For Dummies 0-7645-9803-1
- eBay Business All-in-One Desk Reference For Dummies 0-7645-8438-3
- HDTV For Dummies 0-470-09673-X
- Home Entertainment PCs For Dummies 0-470-05523-5
- MySpace For Dummies 0-470-09529-6
- Search Engine Optimization For Dummies 0-471-97998-8
- Skype For Dummies 0-470-04891-3
- The Internet For Dummies 0-7645-8996-2
- Wiring Your Digital Home For Dummies 0-471-91830-X

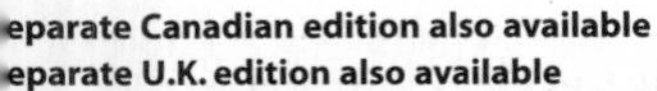

eparate Canadian edition also available

eparate U.K. edition also available

ilable wherever books are sold. For more information or to order direct: U.S. customers visit www.dummies.com or call 1-877-762-2974. . customers visit www.wileyeurope.com or call 0800 243407. Canadian customers visit www.wiley.ca or call 1-800-567-4797.